HEALTHY HOMES

Designing with light and air for sustainability and wellbeing

Nick Baker and Koen Steemers

RIBA Publishing

Published by RIBA Publishing, 66 Portland Place,
London, W1B 1NT

ISBN 9781 85946 713 8

British Library Cataloguing-in-Publication Data

A catalogue record for this book is available from the British Library.

Commissioning Editor: Ginny Mills

Assistant Editor: Clare Holloway

Production: Richard Blackburn

Designed and typeset by Karen Willcox

Printed and bound by Short Run Press

Cover image: Velux A/S

www.ribapublishing.com

FSC www.fsc.org MIX Paper from responsible sources FSC® C014540

HEALTHY HOMES

About the authors

Nick Baker is a course tutor at the Architectural Association and visiting lecturer at the University of Cambridge. His interests include energy modelling, daylight design, natural ventilation and thermal comfort, and occupant-behaviour influences on energy use.

Koen Steemers is Professor of Sustainable Design in the Department of Architecture at the University of Cambridge. An architect and environmental-design specialist, Koen was recently listed as one of the 'Top 50 most influential people in UK sustainability' by *Building Design* magazine.

Acknowledgements

We would like to thank The VELUX Group Knowledge Centre for Daylight, Energy and Indoor Climate for organising a series of expert workshops and report that provided the early stimulus for this book.[1] We are grateful for their support and advice, and for access to their photographic library. We would also like to thank the Martin Centre for Architectural and Urban Studies, Department of Architecture, University of Cambridge, for financial support.

We thank Ben Channon – architect, writer and illustrator – for his engaging illustrations and the trouble he has taken to faithfully convey the technical sense of our hastily scribbled drafts.

Particular gratitude goes to Ginny Mills – commissioning editor of RIBA Publishing – for her expert advice and patient guidance during the preparation of this book.

Finally, in relation to the Care Home scenario in Chapter 6, we would like to thank Michael Parmenter for helpful discussions, and allowing the author access to the care home that he manages.

Dedication

In memory of

Theo C. Steemers

1930–2018

Preface

Our intention in writing this book was to identify the key environmental factors that influence health and wellbeing in housing, elaborate on how and where they interact, and draw conclusions that provide support for design solutions.

Although the intense interest in health and wellbeing in architecture is relatively new, we soon realised that many of the issues – heat and cold, air quality, noise, lighting – have a long history in design literature, for architects, engineers and urban designers. Furthermore, the design expertise developed has always implicitly been directed at health and wellbeing. What new contribution, therefore, could we bring to this topic?

Firstly, we have looked at recent published evidence emerging from a multitude of social and environmental studies. In many areas, modern survey techniques and statistical analyses have made accurate correlations available. For example, no longer do we have to be inspired by a philanthropic hunch that exposure to nature might be good for us – there is now good statistical evidence for this.

Secondly, there are genuinely new areas of research and knowledge that haven't previously touched mainstream environmental design – for example, work on circadian rhythm.

Thirdly, the environment in which we build and live is constantly changing, and not consistently for the better.

There is always a temptation to let the consideration uppermost in our minds dominate others. Thus if we are in 'acoustics mode', we might specify small sealed windows on a façade facing a busy urban street. However, in 'thermal comfort mode' the windows should be openable to facilitate cross-ventilation; in 'view quality mode' they should be bigger; but then again, in 'air quality mode' sealed – against traffic pollution. It is difficult to optimise across various physical parameters, but we have tried to alert the reader to these conflicts, and indicate where quantitative data might be used to resolve a *good enough* solution.

And though some of the measures undoubtedly incur extra costs, many are cost neutral; even if we cannot guarantee effectiveness absolutely, why not *hedge our bets*? Many buildings in which we live and work are not good for us. The Hippocratic Oath, 'Do no harm', reminds us that to eliminate harmful effects, although a modest goal, is a first step.

The book is organised into three parts. Part One reviews the context and historical background, and the growing body of research on health and wellbeing in relation to housing, thus providing the justification for the book[2]. Part Two brings together technical analysis and design synthesis, structured by the homes' *surroundings*, *envelope* and *interior* (including services). Where possible, we have summarised the often-conflicting factors in a simple list of key points.

Finally, in Part Three, we illustrate the process with eight design scenarios. It would have been ideal had these been actual case studies but, as any technical author knows, it is rare to find an example that illustrates all your points precisely and completely. Thus, we have invented plausible examples – often based on real cases.

We hope that this book will provide inspiration and advice, and move the conversation on wellbeing in the built environment forward.

PART ONE:

The background and growing body of research

Chapter 1

The context

Introduction

A house near Paris is to exploit natural light and ventilation well beyond the usual standards. The aim was to deliver an energy-efficient and comfortable home for the Pastours, a 'test' family of four. But it did more than that.

'The house had a positive impact on health. My son had asthma before living in the house, and during the year he didn't have any crises and didn't take any medication. That was the most positive impact that I didn't believe possible ... We spent a wonderful year in that house.' Samantha Pastour, mother of the family residing at Maison Air et Lumière (Figure 1.1).[1]

Seventy years ago, with the establishment of the World Health Organization (WHO), health was defined as 'a state of complete physical, mental and social wellbeing and not merely the absence of disease and infirmity.'[2] In the same year the National Health Service (NHS) was launched in the UK. Today the NHS can barely keep pace with the increasing demands associated with the physical-health challenges of an aging and overweight population, let alone effectively address mental and social wellbeing.

The design of our built environment affects our health and wellbeing, and can have vital implications for quality of life. We spend up to 90% of our time in buildings (about 70% in our homes), so it is inevitable that this environment can create significant opportunities for and threats to our health.

A recent WHO report on housing and health confirms that 'Improved housing conditions can save lives, prevent disease, increase quality of life, reduce poverty ...' and that 'Housing is becoming increasingly important to health due to demographic and climate changes.'[3] The report provides compelling evidence at a global level of health threats ranging from crowding to hazards in housing, but merely lists concerns associated with ventilation and lighting – themes addressed in this book. It provides an excellent overview, policy guidelines, references to standards and research recommendations but relatively little in the way of design-related advice – one purpose of the current volume.

This book addresses key environmental factors that are associated with architectural design and that impinge on health and wellbeing. It does not address health-and-safety hazards, such as those associated with electrical, gas, water or waste systems; or causes of burns, trips and falls.

In this introductory chapter we describe the context with respect to health conditions in homes, the healthcare system and the historical role of medical science in the built environment.

History

There is a long history of science and research related to how our health and wellbeing is affected by housing conditions – most notably, with regard to environmental characteristics. Access to fresh air and sunlight is a recurring theme. Related to this are thermal and humidity conditions, and the spread or triggering of ill health caused by lack of ventilation, poor air quality and overcrowding. More recently research has highlighted housing-design parameters affecting mental wellbeing (discussed in detail in chapter 2). As it is increasingly accepted that mental and physical health are intimately connected,[4] this book addresses both.

Sun worshippers

More than four-and-a-half millennia ago the designs for the great Egyptian cities and pyramids were informed in detail and with astonishing

accuracy by the path of the sun. The sun god Ra was believed to be the source of creation and, after emerging victorious from conquering evil during the night, his morning appearance in the east represented the triumph of life over death. In a most fundamental way, sun and life were linked and embodied in architecture. The Maya civilisation, a thousand years later on a different continent, reflected similar beliefs through the design of its edifices.

Such beliefs and mythologies are underpinned by an evolutionary link between our biology and the rising and setting sun – known today as circadian rhythms – that give cues to diurnal physiological processes. Upsetting these processes can result in ill health (as discussed in more detail later in the book). As we spend approximately 70% of our lives in our homes, ensuring that their architecture embraces the use of natural light continues to be a key design principle to enhance health.

Body and mind

The link between sun and health is not only circadian. In ancient Greece, Hippocrates, the father of medicine, expounded the value of 'heliotherapy' – using the sun to treat patients and support passive, natural recovery (Figure 1.2). Such ideas about health were interpreted in the Roman architectural treatise of Vitruvius, who states,

'The architect should also have a knowledge of the study of medicine on account of the questions of climates, air, the healthiness and unhealthiness of sites ... For without these considerations the healthiness of a dwelling cannot be assured.[5]

Vitruvius goes on to describe, in the context of health, how to orient spaces and design homes in order to create the most favourable conditions.

Figure 1.1: Maison Air et Lumière, Paris.

Four hundred years ago, René Descartes discussed the distinction of mind and body – Cartesian dualism – describing the body as merely mechanical and the mind or soul as irrational, not following the laws of nature. He proposed a two-way link between mind and body – thus, the physical influences the emotional and vice versa.

Today this relationship between mind and body is seen as far more integrated, although in our health service we continue to treat the two as somewhat independent. This is discussed in Jo Marchant's book, *Cure: A journey into the science of mind over body*.[6] The author reveals, through scientific literature as well as case studies, how our physical health (or ill health) is influenced and even determined by a range of interventions and contexts that shape our mental state. These clearly include the built environment and our homes, and the environmental and social conditions that they enable. Marchant presents numerous examples of how the perception of the world around us influences our physical condition. She concludes that

'ignoring subjective experience is great when you're trying to eliminate bias from your scientific experiments but is not always helpful in caring for your patients [i.e. in our case, occupants], *when psychological and physical wellbeing are inextricably intertwined'.*[7]

There is limited conventional medical evidence to address mind-body health effects – Marchant points out that only 0.2% of National Institute for Health Research (NIHR) funding is allocated to this topic – notably, in relation to the built environment. However, there is some acknowledgement from, for example, the *British Medical Journal* of the role of architecture (Figure 1.3), as exemplified in this quote about the design of Maggie's Centres that support cancer patients:

'If any of these buildings contributes to a pleasantly thoughtful or reflective moment for any of its users, a moment with friends or relatives, or a moment of hope and calm that they otherwise wouldn't have had then they have already achieved something wonderful.'[8]

Subjective and objective experiences of our homes both impact on our mental wellbeing and physical health. Much of the scientific research has historically focused on physical parameters and physiological health (as outlined in the brief overview below). However, increasingly there is evidence that the qualitative and social conditions that our built environment creates also affect our mental wellbeing.

The development of health science

This section focuses on the developments of medical research in relation to domestic architecture in the 19th and 20th centuries. The history of sanitation (particularly regarding sewage) is a fascinating story that highlights the rising dominance of medical specialists in domestic design. This story is expertly relayed by Dr Annmarie Adams in her book *Architecture in the Family Way*[9] (notably, the chapter 'Doctors as Architects'), in which she states,

'The consequences [of the dominance of the medical specialist in domestic design] *for the architectural profession in England were devastating. A "systematic" view of the house professed by the medical doctors was embraced by the middle class, with the result that there was a widespread wariness of building professionals. Design based on scientific principles ... informed much material culture of the period.'*[10]

Figure 1.2: The Greek notion of heliotherapy suggests that letting sunlight into our homes has health benefits.

Adams goes on to add the following:

'Many of those [physicians] *associated with the Domestic Sanitation Movement became, in a sense, "building-doctors", called on to diagnose, treat, and heal architecture, much as they might a patient.'*[11]

The current section does not reprise this history of sanitation, but provides some highlights of the development of medical research in relation to broader design considerations of fresh air and good lighting.

Sunlight

Inspired by the writings of ancient thinkers such as Vitruvius, the medic Bernhard Christoph Faust (1755–1842) developed his solar theory related to urban planning – *Sonnenbaulehre*. Through the orientation and arrangement of buildings he sought to ensure increased access to sunlight and air flow in order to create healthy living conditions. He also considered the social benefits of street access, gardens and playgrounds. By 1825, the Munich architect, Gustav Vorherr (1778–1847), recognised Faust's expertise and had put his principles into practice in planning the aptly named *Sonnenstrasse* (Sun Street) in the centre of that city.

The theory of phototherapy, harking back to Hippocrates' heliotherapy, was further refined by Niels Ryberg Finsen (1860–1904), who demonstrated that specific wavelengths of light (notably ultraviolet – UV) can be shown to have medical benefits. As the sun is the most abundant natural source of UV light, Finsen's research provided the scientific evidence for the benefits of solar access. He won the Nobel Prize in 1903 for his research on the use of light radiation for the treatment of diseases, such a *lupus vulgaris* (a form of tuberculosis – TB – that attacks the skin). Light therapy became widespread and influenced planning and building design, even beyond the introduction of TB vaccination in the 1950s.

Fresh air

Despite these solar-design initiatives, poorly ventilated and overcrowded housing was, by the start of the 19th century, the living condition experienced by the urban poor. Combined with bad sanitation (found also in middle-class homes of the time) and malnutrition, this resulted in the dramatic rise of TB, an infectious disease of the lungs. TB, at its peak in the first half of the 19th century, was claiming the life of one person in four in Europe and North America. It took nearly a further century for medical drugs to be discovered and administered in widespread public inoculation programmes. However, in the meantime there was a decline in the disease as the result of public-health policy related to improving housing conditions. In the UK this culminated with the Housing Act 1930, which spelled the end of 'back-to-back' housing – a building form that prevents cross ventilation and limits access to daylight.

An early building-design response to TB was the Sanatorium Movement, initiated in 1854 by Hermann Brehmer (1826–1889), a physician who had recovered from TB after a period of time spent in the Himalayas. Access to clean air was a key aspect of the treatment, and patients would spend time outdoors to recuperate. Open-air verandas, terraces or balconies were common features of recovery spaces, and sometimes included rotating pavilions that could follow the path of the sun.[12] By the end of the century, the medical establishment, initially sceptical, accepted the value of sanatoria to treat TB. Sunlight and the open air would become the mantra for subsequent sanatoria projects and subsequent modern building design.

Despite the advances in knowledge and vaccination programmes, TB is today still one of the top 10 causes of deaths worldwide. The WHO states that '[i]n 2017, 10 million people fell ill with TB, and 1.6 million died from the disease' and that therefore '[e]nding the TB epidemic by 2030 is among the health targets of the Sustainable Development Goals'.[13] The strategies to end TB continue to recognise that poorly ventilated and crowded homes are a key risk factor.

Figure 1.3: Creating settings for a moment of calm or a moment with a relative can 'achieve something wonderful'.

Carbon dioxide

The chemist and innovative hygienist, Max Pettenkofer (1818–1901), was known for his work related to public-health aspects of sewage, clean water and fresh air. He studied natural air exchange and artificial ventilation in numerous buildings, and employed a new method of measuring carbon dioxide (CO_2) to determine the air's purity. In 1858 he published a paper entitled 'Über den Luftwechsel in Wohngebäuden'[14] (Concerning Air Exchange in Residential Buildings), in which he recommended a maximum concentration of 0.1% of CO_2 by volume of air in occupied rooms, or 1,000 parts per million (ppm) – known as the Pettenkofer number. This threshold value is still widely used today.

Environmental theory

Florence Nightingale (1820–1910), the founder of modern nursing who used statistical analysis to provide evidence for her findings on healthy environments, wrote,

'There are five essential points in securing the health of houses:–

1. *Pure air.*
2. *Pure water.*
3. *Efficient drainage.*
4. *Cleanliness.*
5. *Light.*

Without these, no house can be healthy.'[15]

She went on to add other environmental factors that support health, including warmth, sound, variety, colour, form, views and diet. These are still very pertinent today and are addressed in this book. However, she reserved her most emphatic statements for fresh air and light:

'The very first canon of nursing, the first and the last thing upon which a nurse's attention must be fixed, the first essential to a patient, without which all the rest you can do for him is as nothing, with which I had almost said you may leave all the rest alone, is this: TO KEEP THE AIR HE BREATHES AS PURE AS THE EXTERNAL AIR, WITHOUT CHILLING HIM.'

'It is the unqualified result of all my experience with the sick, that second only to their need of fresh air is their need of light; that, after a close[d] room, what hurts them most is a dark room. And that it is not only light but direct sun-light they want.'

Nightingale's understanding represents one of the more comprehensive and convincing theories of its time, of the relationship between the built environment and health.

Figure 1.4: Letchworth Garden City, Meadow Way Green (1900) by Parker and Unwin.

Planning and design

The accumulation of evidence and knowledge, as outlined above, was to impinge on important urban and architectural initiatives. The Garden City Movement, founded in 1898 by Ebenezer Howard (1850–1929), had as an aim '... to raise the standard of health and comfort of all true workers of whatever grade – the means by which these objects are to be achieved being a healthy, natural, and economic combination of town and country life ...'[16] The intended integration of nature and the limited density of development offered access to fresh air and sunlight. The movement was a direct reaction to the overcrowded and unsanitary conditions of cities at the time. The Garden City ideals were realised in practice in the UK by Barry Parker (1867–1947) and Raymond Unwin (1863–1940), architects of the world's first Garden City at Letchworth (1903) (Figure 1.4). The Garden City Movement had a global influence on planning. Its principles still underpin New Urbanism and Eco-towns, except with respect to increasing urban density, to counteract car use and to make neighbourhoods walkable.

As the architect of Letchworth, Parker was aware of health design strategies that went beyond general planning to more detailed house design. Writing about the influence of TB on Modernism, Margaret Campbell made the following observation:

'In 1914, Barry Parker (1867–1947), the Garden City Movement architect, had a sleeping porch constructed to the top floor of his Letchworth house, Crabby Corner. A contemporary photograph of the interior shows a single bed

Figure 1.5: Zonnestraal Sanatorium (1926–8) by Johannes Duiker.

placed directly under the open corner, but it was also intended that one could sleep out in the open air. This feature was similar to the open-air sleeping cabins requested for the Californian beach house at Newport, designed in 1922 by the modernist American-Austrian architect Rudolf Schindler for the health-fanatic Dr Philip Lovell.[17]

We have already mentioned sanatoria, and by the start of the 20th century their architecture embodied the healing principles of clean air and sunlight as the only available treatment for widespread TB. Buildings like Johannes Duiker's (1890-1935) Zonnestraal (sunbeam) Sanatorium (1926–8) in the Netherlands (Figure 1.5) and Alvar Aalto's (1898-1976) Paimio Sanatorium (1929–33) in Finland (Figure 1.6) underpinned the principles of emerging Modernist and healthy architecture: rational, open, light, white, transparent, clean, etc. Although there had been earlier precedents (if less architecturally emphatic) for teaching in the open air, Duiker went on to apply the same architectural notions to an educational building, the Open Air School (1929–30) in Amsterdam. This was followed five years later by Eugène Beaudouin's (1898-1983) and Marcel Lods' (1891-1978) École de plein air in Suresnes near Paris, with its distinctive 'concertina' window screens.

At an urban residential scale, a group of Modernist architects, led by Ludwig Mies van der Rohe (1886-1969), designed the Weissenhofsieldung (loosely translated as 'white courtyard settlement') in Stuttgart, Germany in 1927 (Figure 1.7). The intentions of the settlement were to demonstrate efficiency, using materials and technologies to lower costs, 'gaining the maximum effect for the minimum amount of means', leading to 'improvements of living itself'.[18] The white, flat-roofed, mixed housing development has become an influential icon of Modernist housing, with designs that can be linked back to TB-treatment principles.[19]

The first Modernist house to have an explicit health agenda was arguably Richard Neutra's (1892-1970) Lovell House (1927–9) in Los Angeles, sometimes referred to as the Health House (Figure 1.8). The client, Dr Lovell, has already been described, above, as a 'health-fanatic' and was patron to like-minded Modernist architects like Rudolf Schindler (1887-1953) and Neutra, so it is perhaps no surprise that this residence combined the concepts of health and Modernism. Although Neutra's ideas went beyond physical health to the (unsubstantiated) claim that his design could cure neuroses,[20] it is the Modernist tropes that are most evident: white walls, glazed façades, ribbon windows and southerly orientation – with patios, flat roofs and balconies providing indoor/outdoor sleeping and exercise spaces, including a swimming pool.

Energy efficiency

The energy crisis of 1973, caused by an oil embargo by the Organization of the Petroleum Exporting Countries (OPEC), resulted in a quadrupling of energy prices. In terms of the construction sector this led to initiatives to reduce energy demand and the incorporation of renewable energy sources. In the UK, the main energy demand in housing is for heating, due to heat losses through the building envelope. By 1975 thermal transmittance values (U-values) had been reduced (from 1.7 to 1.0 watts per square metre degrees Kelvin [W/m^2K] for walls, and from 1.4 to 0.6 W/m^2K for roofs), and by 1990 window areas were limited to 15% of the floor area. The generously glazed Modernist house typology was deemed to be energy wasteful.

The long-term response to the energy crisis resulted in increased airtightness and reduced window sizes – the opposite of 'open air and light'. These strategies have inadvertently led to

Figure 1.6: Paimio Sanatorium (1929–33) by Alvar Aalto.

Figure 1.7: Am Weissenhof 14–20 (1927), Weissenhofsiedlung by Ludwig Mies van der Rohe.

a reduction in natural light, limiting of views and reduced indoor air quality (IAQ). In the meantime, the phenomena of Seasonally Affective Disorder (SAD) and Sick Building Syndrome emerged, causing occupants to suffer from symptoms of ill health or feeling unwell, associated with reduced availability of light, poor electric lighting, a lack of fresh air and internal contaminants. These were the unintended consequences of focusing too narrowly on reducing energy demand. There is an urgent need to link energy efficiency with health in housing design – a key purpose of this book.

Circadian science

A number of 'chronobiologists' – such as George C. Brainard[21] and Russell Foster[22] – are currently working on the relationships between the physiology of the eye, light and circadian regulation. The discovery in 2000 of light-sensitive ganglion cells in the eye that are not associated with vision but do influence circadian regulation has added further impetus to the field.[23] Circadian rhythms are modulated by the diurnal light cycle, and influence sleep patterns, brain activity, metabolism, hormonal production and behaviour. It is now well known that disruption of circadian regulation is linked with a wide range of negative health consequences, including 'increased risk [of] premature death, cancer, metabolic syndrome, cardiovascular dysfunction, immune dysregulation, reproductive problems, mood disorders, and learning deficits'.[24] We have evolved as an outdoor species, our lives following the daily light-dark cycle, but now we spend the vast majority of our time inside, largely deprived of sunlight. It is no surprise that our health can be affected, particularly in urban environments and in homes where daylight is scarce and where at night we are exposed to electric lighting, which can disrupt circadian regulation. An understanding of the timing of exposure to light (e.g. morning or evening), the amount of light received (e.g. intense direct sunlight or artificial light) and the colour of light (e.g. cool blue or warm red) is now considered valuable in informing the way we analyse daylight in architecture[25] and capable of informing healthier housing design – as discussed in chapters 4 and 5 of this book.

Data and standards

The statutory standards with which almost all British homes must comply are those of the UK Building Regulations. Although progressively tightened since the OPEC oil crisis, these still lag behind other, voluntary, measures that are more progressive in terms of energy use and health. One of the earliest of these packages was 'Lifetime Homes'. Developed in the early 1990s by

Figure 1.8: Lovell Health House, Los Angeles (1927–9) by Richard Neutra.

housing experts including the Joseph Rowntree Foundation, these dwellings aim to support the changing needs of individuals and families at different stages of their lives (see 'Moving and access' in chapter 2).

The US-originated LEED (Leadership in Energy and Environmental Design) 'green'-building rating system is also being increasingly widely used (see 'Building contents, construction materials and finishes' in chapter 5 and the 'New-build prefabricated mass housing' example in chapter 6). Other UK-based measures more tightly focused on health and wellbeing are listed below.

Healthy Homes Barometer

For a number of recent consecutive years the 'Healthy Homes Barometer' has used survey data of between 14,000 to 250,000 European adults to assess the relationships between health and housing.[26] Early results revealed that Europeans identified their home as being the most important context for positive health. Sleep, ventilation and daylight ranked higher in importance for health than smoking, exercise and diet. Despite this, occupants did little to improve conditions in the short (e.g. increased window opening for fresh air) or longer term (e.g. renovation, typically instigated due to space needs or decorative conditions).

Subsequently the Healthy Homes Barometer confirmed previous findings that sleep, fresh air and daylight were the key characteristics of healthy living. However, new additions to this list were thermal comfort and appropriate humidity, in the following order and with the following findings:[27]

- Sleep: Occupants in homes where they report getting a good night's sleep are 50% more likely to report good health and feeling energised.
- Comfort: Cold homes correlate with a 50% increase in reported nose and throat infections.
- Air: Those occupants that do not air their homes complain twice as much as others about a lack of feeling energised.
- Light: Conversely, having enough daylight halves the risk of feeling a lack of energy.
- Humidity: Where mould is perceptible, there is a 50% increase in reported throat infections.

Cumulatively, these health implications have a large effect not only on overall public health and wellbeing but also on absenteeism at work and resultant economic costs. These are discussed further in the section, 'Pressures on healthcare', later in this chapter.

The headline findings from a recent Healthy Homes Barometer adds further insight to the impact of buildings on health.[28] One in six Europeans live in homes that are damp, poorly daylit or thermally uncomfortable. Overall, more than 1½ times as many occupants living in such unhealthy homes report poor health compared with those in healthy homes. Damp or mould alone increases the likelihood of asthma by 40%. In households with fuel poverty (i.e. unable to keep the home warm, and associated damp problems), twice as many occupants report suffering poor health.

Renovation has significant potential. Homeowners can be incentivised to renovate their homes equally highly for energy, comfort and health improvements, with three out of four stating that each of these would motivate them.

Green Building Council

The increasing consumer awareness of, and interest in, health and wellbeing in housing has been recognised in a recent UK review. The UK

Green Building Council (UKGBC) published an influential report entitled 'Health and Wellbeing in Homes.'[29] This not only identified existing research, and placed health and wellbeing firmly on the housing design agenda, but also presented market research to identify the business case. Headline findings include the fact that almost 30% of homeowners and renters would be willing to pay more for a healthy home. This is supported by reported features of an ideal home, which rank safety, low maintenance, no condensation, views of nature, affordable warmth, good daylight and reduced noise above financial concerns related to the value of the home.

Home Quality Mark (BRE)

The Home Quality Mark (HQM) has been developed since 2016 by the UK's Building Research Establishment (BRE) as part of its suite of BRE Environmental Assessment Method (BREEAM), and currently only applies to *new* homes.[30] The Mark – focused on a number of environmental, technical and economic performance criteria – also recognises the impact on health and wellbeing. Scoring for health and wellbeing takes account of indoor air, temperature, light and noise but also access to local amenities. The HQM sets minimum required thresholds (typically based on the current Building Regulations or similar standards) across all the criteria. There is an overall score (out of 500 credits) that can be expressed simply as a 5-star rating. For the consumer, the Mark breaks down the scoring (and relative weighting) into three sectors – cost (28%), wellbeing (30%) and footprint (42%) – so that it is possible to identify health-related performance independently from economic and environmental issues. The technical manual cites mainly standards, regulations and policy documents as opposed to direct source material from scientific literature. The development of the Mark has relied on expert consultation and advice. To date, over 10,000 homes have been registered with the HQM – although, unlike the WELL Standard (discussed below), these are not searchable.

WELL Building Standard

The International WELL Building Institute launched the WELL Building Standard, after six years of research and development.[31] Unlike the HQM, WELL focuses specifically on health and collates a wealth of scientific research to underpin seven core concepts related to health: air, water, nourishment, light, fitness, comfort and mind. These are reminiscent of the principles established by Florence Nightingale over 150 years earlier (see 'Environmental Theory' earlier in this chapter). The research is distilled into a comprehensive and detailed checklist of performance criteria, and has similar strengths and weaknesses to other assessment methods, including the HQM: criteria are applied after the scheme-design stage, are tactical rather than strategic, focus more on quantitative than qualitative considerations, tend to be global rather than responding to specific local conditions, etc. Nevertheless, both the HQM and the WELL Standard are excellent drivers for change, awareness raising and stakeholder engagement.

The WELL Standard is primarily focused on work environments (only 143, or 15% of, registered projects are residential and can be searched), but includes new-build and refurbishment. It misses out to a large extent on one of the WHO's three primary areas defining health (physical, mental and social wellbeing) – namely, the social, which is particularly relevant for housing. Nevertheless, it marks a significant shift in awareness of the potential impact of architecture on health and wellbeing. The strength of such methods is a focus on a top-down application of criteria, but there has also been an increasing amount of bottom-up evidence of the impact of housing on health and wellbeing.

An improved understanding (whether from data or assessment methods) with regard to occupant health and wellbeing, when integrated into design strategies and standards, should result in healthy homes that need not cost more. Good design and cost-effective measures can lead to improved health outcomes. At the societal level it makes financial and moral sense to implement such measures, which can lead to savings to the healthcare system, improved productivity (lack of absenteeism) health equity and a higher quality of life (Figure 1.9). Such considerations are introduced in more detail below.

Figure 1.9: Comfort, fresh air and light rank high in occupant perception of health and wellbeing in homes.

Pressures on healthcare

The healthcare system is facing enormous challenges, and anything that can be done to prevent ill health, help the infirm and support recovery through the design of our homes will contribute to improving the situation. The pressures are a result of a number of coinciding changes,[32] some of which can be addressed through the design of homes and neighbourhoods:

Lifestyle factors: alcohol consumption, smoking, poor diet, insufficient exercise – the last two in particular are linked with domestic and neighbourhood design (see chapter 2).

An ageing population: living longer with illnesses such as diabetes, heart and kidney disease – design to Lifetime Homes standards and make neighbourhoods socially integrated and accessible (see chapter 2).

Changing public expectations: advice on managing our health, dealing with mental health and social care, more efficient appointments, processing and access to medication – design for social and mental wellbeing (see chapter 2), integrate digital health technologies in the home and provide local health and wellbeing support.

Avoidable pressures on A&E: due to lack of access to General Practitioners (GPs), poor health management, winter crises – improve domestic health and thermal performance, including affordable warmth and avoidance of summer overheating (see chapter 5), as well as integrating neighbourhood and digital health opportunities.

> *Increasing costs:* of new treatments and medicines but also of services and energy, leading to a potential funding gap of £30 billion by 2020 – incorporate energy efficiency (see chapter 4) and preventative health and wellbeing strategies (described throughout this book).

The financial challenges for the healthcare system are widely recognised.[33, 34] The influential Marmot Review identified the economic cost of inequality in health:

'It is estimated that inequality in illness accounts for productivity losses of £31–33 billion per year, lost taxes and higher welfare payments in the range of £20–32 billion per year, and additional NHS healthcare costs associated with inequality [which] *are well in excess of £5.5 billion per year. If no action is taken, the cost of treating the various illnesses that result from inequalities in the level of obesity alone will rise from £2 billion per year to nearly £5 billion per year in 2025.'*[35]

The direct annual cost to the NHS of poor housing has been estimated to be £2.5 billion.[36] Cold homes alone are calculated to cost the NHS in England £1.36 billion a year.[37] For every £1 invested in reducing fuel poverty, the NHS will reap 42 pence in savings (not to mention job creation and tax revenues, reduced reliance on imported energy and increased economic growth as a result of energy-efficiency policies – so that the total financial benefit would rise to £1.27 for every £1 invested by 2030).[38]

Even relatively minor ill health that is in part attributable to unhealthy housing conditions – such as coughs, colds, 'flu and sickness – results in the UK in 27 million lost working days and an estimated cost of £1.8 billion annually. This does not include a further 15 million workdays lost per year due to mental ill health such as stress, depression and anxiety. The total annual cost to the economy of absenteeism through ill health is estimated to be £11.6 billion.[39]

A recent report concludes that the best strategy to improve the situation, with respect to patient care in particular, is to strengthen community-based services and focus on prevention – both of which present roles for the design of housing and neighbourhoods.[40] There is now clear recognition that the social and housing contexts must not be overlooked as determinants of ill health (Figure 1.10):

'The circumstances in which people live and work are as important for communicable as they are for non-communicable disease. Social conditions powerfully influence both the onset and response to treatment of the major infectious diseases that kill.'[41]

More than simple economic cost and physical health, Marmot and his committee recognise that wellbeing and sustainability are important societal goals. Although not explicitly discussed in the report, three of the six key policies identified by Marmot can be related to homes as a focus to change health inequalities:

1. Ensure healthy standards of living for all – here, Marmot concentrates on income, not on housing standards per se.
2. Create and develop healthy and sustainable places and communities – Marmot emphasises sustainable neighbourhood design and energy efficiency (notably fuel poverty), omitting housing-design considerations and their more direct role for health and wellbeing.
3. Strengthen the role and impact of ill-health prevention – Marmot proposes a focus on ill-health prevention programmes, not on the potential for ill-health prevention through design.

Figure 1.10: Creating positive circumstances and opportunity for social interaction where people live influences the onset and response to treatment of ill health.

The indicators of health that Marmot proposes are:[42]

- life expectancy – to capture years of life
- health expectancy – to capture the quality of those years
- wellbeing – Marmot is unable to define an indicator of wellbeing as the third national target on health inequality

Health experts have found it challenging to identify evidence linking housing-design strategies with physical health and mental wellbeing. However, as Marmot notes, there is clear recognition that poor housing and neighbourhood conditions are very significant:

'Bad housing conditions – including homelessness, temporary accommodation, overcrowding, insecurity, and housing in poor physical condition – constitute a risk to health. A study carried out by Shelter in 2006 suggested that children in bad housing conditions are more likely to have mental health problems, such as anxiety and depression, to contract meningitis, have respiratory problems, experience long-term ill health and disability, experience slow physical growth and have delayed cognitive development. These adverse outcomes reflect both the direct impact of the housing and the associated material deprivation.'[43]

The UKGBC goes further than simply identifying that poor housing conditions are a cause of ill health, and presents a plethora of design factors and diverse evidence for their potential roles in determining healthy homes.[44] The challenge remains how to prioritise, quantify and integrate numerous and potentially conflicting characteristics. This book addresses how housing design can avoid adding to the challenges (do no harm), improve the situation (help) and encourage behaviour (self-help) that leads to better health and wellbeing outcomes.

'Nudge' architecture

The publication of *Nudge: Improving health, wealth and happiness* by Richard Thaler and Cass Sunstein in 2008 was influential in revealing that behaviour can be strongly influenced by context.[45] People can be nudged into making better decisions in largely automatic, non-coercive and simple ways, through changing what Thaler and Sunstein refer to as 'choice architecture'. Can actual architecture create the metaphorical choice architecture? The role that it can play seems evident: 'Design-led interventions can make better choices easier or constrain behaviours by making certain actions more difficult.'[46]

There can be no doubt that negative physical health-related considerations associated with, for example, poor indoor environmental quality should be avoided. However, there is an equal need to focus on supporting positive mental wellbeing, which in turn has implications for and links with physiological health. There is an established body of expertise related to the study of physical health with increasing quantitative evidence, but research into wellbeing in the built environment is a relatively recent and hitherto qualitative area of investigation that is nevertheless beginning to reveal consistent and widely accepted findings.

When we discuss health and wellbeing in buildings, it is important to incorporate a wide range of both quantitative and qualitative considerations rather than focus on single, narrowly defined criteria. Such 'silo thinking' tends not to aid good design and often different criteria are in tension. In addition, perfectionism can be crippling – or, to paraphrase Voltaire, 'the best is the enemy of the good'. An alternative approach is to determine 'good enough' strategies, which increase diversity and adaptability and are user-centred. This is not to deny the potentially chronic health impacts of poor indoor environmental quality on certain sectors of the population (i.e. large impact for a small population), but rather to balance and complement this with strategies to improve health and wellbeing for the wider public (i.e. improvements for a large population).

Summary

Since industrialisation and the move of populations from an agrarian to an urban environment during the 19th and 20th centuries, humankind has increasingly become an indoor species. However, in evolutionary terms we are an outdoor species. The fact that we now spend 90% of our time inside has had consequences for our health. These consequences are now more fully appreciated as reflected in new data, surveys of occupants and emerging standards related to health in buildings.

At the same time, changing demographics, behaviour and expectations are putting new pressures on our health services. The overall societal financial burden of poor health in the UK is estimated to be 60% of the annual national health budget (ie £75 billion - based on the research cited in this chapter - of the circa £125 billion total NHS budget). Three of the six policies proposed by the Marmot Review to address health challenges are directly associated with or situated in the domestic environment. Thus, housing design has a fundamental role to play in the nation's health.

Health is not just about the treatment of physical ill-health but also relates to mental and social wellbeing. Housing design has the potential to support behaviours that positively contribute to health as defined in these more holistic terms, although there has been and continues to be a focus on individual concerns. Such a focus can lead to unintended consequences – so rather than optimising for individual characteristics, we propose an approach that offers diversity and adaptability to suit a range of needs.

In the 19th century the history of healthy homes reveals an interesting tension between the role of the architect and the medic. It was the medical specialist, the 'house-doctors', who tended to determine the necessary requirements – particularly in relation to ventilation, drainage or light. Architects were somewhat marginalised from the debate. However, by the 20th century, the understanding of design strategies to improve health and wellbeing became established and adopted in architectural design and planning. In the meantime, medical research continued to develop drugs and vaccination programmes to deal with major diseases such as TB. The interchange between health specialists and building professionals continues today, and lies at the heart of the discussion in the next chapter.

Chapter 2

Physical and psychological health

Introduction

In the field of sustainable development, reference is often made to the 'triple bottom line' of physical, economic and social benefits. The health and wellbeing triple bottom line could be summarised as health, comfort and happiness. In order to draw more direct parallels with the built environment, we can refer to Vitruvius and his tripartite model of the three elements required for a well-designed building (Figure 2.1):[1]

- *firmitas* or firmness (health)
- *utilitas* or commodity (comfort)
- *venustas* or delight (happiness).

In this context health is referred to in more conventional terms, as the absence of disease, and typically measurable in terms of symptoms such as body temperature or blood chemistry. Comfort is widely understood to be a 'condition of mind which expresses satisfaction' with the environment[2] – whether thermal, visual, acoustic or otherwise – and thus incorporates both qualitative psychological considerations (e.g. expectation, control) and quantitative physical parameters (e.g. temperature, air movement). Happiness colloquially refers to emotions experienced, potentially ranging from contentment to joy. Happiness is therefore primarily a subjective consideration. Despite this, research over the last decade has begun to quantify wellbeing; this will be addressed in more detail in this chapter.

Defining health and wellbeing

One key challenge is the quantification of health and wellbeing, and thus the assessment of the overall health performance of a design. At one end of the spectrum, physical ill health is typically identifiable and measurable in terms of symptoms and causes. For example, air quality – e.g. Volatile Organic Compounds (VOCs), Particulate Matter (PM) or CO_2 – and its impact particularly on vulnerable occupants (e.g. those with lung conditions, the young and the old) can be quantified, and treatments of both occupants and buildings can even be prescribed (e.g. improved ventilation, the removal of offending materials, design interventions to avoid mould growth).

Figure 2.1: Health, comfort and happiness are interconnected characteristics of wellbeing, similar to the Vitruvian tripartite diagram (left), but they also lie on a spectrum from the absence of physical ill-health to the presence of mental flourishing (right).

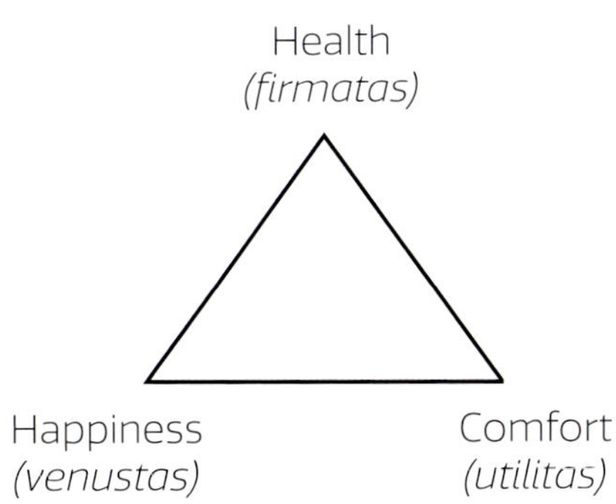

Although subjective assessment of air quality, particularly related to odour, can offer useful insights, health-threatening indicators are often impossible to sense and can only be measured (e.g. carbon monoxide). Specific criteria and design strategies to tackle chronic physiological health problems can be defined, and there is a wealth of expertise to support this.[3]

At the other end of the health-and-wellbeing spectrum is mental wellbeing or happiness. As we move along the spectrum from objective physical health to subjective psychological health, the common perception is that the emphasis changes from quantitative to qualitative. However, it is now evident that even within the sphere of subjective parameters there are methodologies and indicators that can be defined. For example, the field of thermal comfort has seen a development from narrow and precise physiological comfort theory, based on the seminal work of Povl Ole Fanger,[4] to a more holistic understanding that has led to the acceptance of adaptive-comfort theory.[5] Similarly, health research has extended from the treatment of symptoms to incorporate a wider and more holistic appreciation of wellbeing of the population. It is this that forms the primary focus of this chapter.

The notion of wellbeing consists of two key elements: feeling good and functioning well. Feelings of happiness, curiosity and engagement are characteristic of someone with a positive sense of themselves. Having positive relationships, control over your own life and a sense of purpose are all attributes of functioning well. International evidence has recently been gathered to measure wellbeing, demonstrating that this field has now emerged as a rigorous discipline.[6]

Recent research has demonstrated the link between key design characteristics of the built environment with the 'Five Ways to Wellbeing' (Connect, Keep Active, Take Notice, Keep Learning and Give), which are associated with positive mental health.[7] Based on such findings, the following paragraphs reveal how the provision of local urban and domestic resources can impinge on the five healthy behaviours. This supports current theory and research, which shows that a sufficient quantity and quality of diverse environmental, social and physical resources in the built environment can influence human cognition – which in turn can increase healthy behaviours in the wider population.

The 'Five Ways to Wellbeing'

The relationship between architecture and health has received limited attention beyond the specific design requirements of health buildings. Recent work has changed this, and has established a more holistic awareness of the role of architecture in health. An example of this in the UK includes the publication of reports by the Royal Institute of British Architects (RIBA)[8] and the Commission for Architecture and the Built Environment (CABE).[9] This is supported by an increasing wealth of medical research related to physical health[10] and mental health.[11] The emphasis has been on ill health as a result of the effects of environmental characteristics such as overcrowding, noise, air quality and light. These effects are typically described as direct (i.e. consequences on physical and mental health) or indirect (e.g. through social mechanisms).[12] However, rather than focusing on ill-health, the definition and study of wellbeing has been emphasising the behaviours that support a 'flourishing' population. It is the built-environment characteristics that support such positive behaviour that form a key point of discussion here.

The science of wellbeing is a relatively recent area of enquiry. However, the UK Government's 'Foresight' project related to wellbeing[13] provided the critical mass of evidence that led to the definition of the 'Five Ways to Wellbeing' mentioned above.[14] These represent the key behaviours that have been shown to relate to improved wellbeing. Each is associated with subjective wellbeing as reported in research papers, notably in medical journals, that draw on large-scale studies and meta-analysis of exacting studies.[15] Thus, there is no shortage of evidence to support the assertion that the 'Five Ways' result in improved wellbeing.

1 Connect: The quantity and quality of social connections (e.g. talking and listening to family or strangers) correlates with reported wellbeing as well as physical health.[16]
2 Keep Active: There is ample evidence from global studies and meta-studies[17] to demonstrate that physical activity reduces symptoms of mental and physical ill health.[18]
3 Take Notice: Being mindful – paying attention to the present and being aware of thoughts and feelings – is a behaviour that reduces symptoms of stress, anxiety and depression.[19]
4 Keep Learning: Aspirations are shaped in early life, and those who have higher aspirations tend to have better outcomes. Such aspirations are modified by the environment.[20] The evidence shows that, also later in life, those participating in, for example, music, arts and evening classes attain higher subjective wellbeing.[21]
5 Give: Evidence has emerged that pro-social rather than self-centred behaviour has a positive impact on happiness. Such consequences of altruistic behaviour are related both to spending on others as opposed to oneself,[22] and volunteering and offering help.[23]

The critical next question is to discuss how the 'Five Ways to Wellbeing' relate to and are influenced by the built environment.

Connect

The provision of local 'everyday public spaces' creates opportunities for people to connect, and is a significant resource of wellbeing for individuals or the wider community.[24] Although not all users have the same requirements and expectations of a social space, key qualities include:

- location – accessibility and proximity to other communal resources (school, market) to support casual encounters
- places to stop and sit, on a park bench or at a café table, so that encounters can be more than fleeting
- adaptability – spaces without specific or prescribed functions, which enable spontaneous, impromptu activities
- homeliness – a sense of safety and familiarity
- pleasantness – clean and peaceful, or bustling and lively
- specialness – unique qualities, aesthetics or subjective memories.

When a space is pedestrian-oriented as opposed to car-oriented, this correlates with a sense of community due to the perception of the pedestrian environment being particularly strongly related to opportunities for social interaction.[25] And finally, natural, green or landscape qualities have been widely and for a long time associated with a range of health benefits.[26] In summary, 'public spaces that brought people together and where friendships and support networks were made and maintained were key to a general sense of wellbeing'.[27] (Figure 2.3)

Keep Active

Physical activity (walking, cycling, sports, etc.) is widely associated with reducing causes of chronic conditions and the burden of disease, disability

Figure 2.2: 'Keep active': Creating stairs with views, daylight and interest is more likely to encourage movement, for fun or as a more enjoyable connection between key parts of the home.

and premature death. Design characteristics associated with increasing activity include:

- access to physical activity facilities (e.g. sports centres and equipment)
- convenient and proximate access to destinations (work, shops, school, public transport)
- high residential density (which is associated with greater proximity to facilities and destinations)
- land use – e.g. mixed use – and walkability (convenient and safe pavements, traffic-calming features).[28]

Although there are some potential additional benefits to physical activity in an outdoor and preferably natural environment, exercise indoors can be effective[29] (Figure 2.2). Design strategies to promote indoor physical activity include:

- the provision of (shared) exercise space
- encouraging stair use through the distribution (separation) of functions over different floor levels
- creating attractive experiences along circulation routes (views, art, daylight, greenery).

Figure 2.3 'Connect': Pedestrian, safe and green spaces offer opportunities for social interaction, supporting a general sense of wellbeing.

Take Notice

Being mindful and taking notice with respect to an architectural or urban-design context is a behaviour for which there is only recent evidence. However, in a randomised control test, the provision of art, planting and landscaping, wildlife features (e.g. insect boxes) and seating are examples of the kind of interventions that have resulted in significantly increased observations of people stopping to take notice.[30] The same study also showed that diverse types of open space (combining soft with hard landscaping) and a higher relative proportion of public to private space is also associated with increased reported mindfulness (Figure 2.4).

With regard to Take Notice's aforementioned contribution to anxiety relief, designers could take a cue from Swedish Modernist Erik Gunnar Asplund: in his 1936 extension to Gothenburg's classical Law Courts building, his single large staircase leading up to the courtrooms has steps so deliberately shallow that the visitor (often under considerable stress before a court hearing) is forcibly slowed down and calmed by the simple experience of walking up them.

Keep Learning

There is evidence from educational research that the physical environment of the home and classroom are mediating variables that influence intellectual development. Domestic parameters include a home that is clean and uncluttered, appears safe for play and is not dark or monotonous (Figure 2.5).[31] The distance and orientation of seating in relation to others will

Figure 2.4: 'Take notice': Mindfulness is encouraged by the availability of planting – even pot plants on a balcony.

Figure 2.5: 'Keep learning': Creating diverse environments to support various ways of learning is one design response to improving wellbeing.

influence the level of interaction and dialogue. For example, people facing each other will converse more readily than those adjacent to each other. Unobstructed eye contact is an important variable, particularly in an educational context, making a semicircle classroom-seating arrangement most effective.[32] At a more prosaic level, in order to support learning, interior environments need to be physically and thermally comfortable, safe, well lit, quiet and have clean air. However, there is evidence that learning will improve when comparing a poor environment (a run-down and poorly maintained space) with an adequate one (i.e. one that is 'good enough'), but that further and more extravagant facilities (e.g. specialised spaces or digital equipment) do not show further improvements in learning.[33] As previously mentioned, the opportunity to engage in art, music and evening classes increases wellbeing, and thus such activities should be accommodated in the design of homes (e.g. light, cleanable spaces for art, soundproof spaces for music) and neighbourhoods (e.g. local communal spaces for classes).

Give

The presence of environmental stressors reduces helping behaviour. Beyond the broad evidence outlined above related to other ways to wellbeing, there is research that connects the physical environment with neighbourhood social capital.[34] The fact that people are less altruistic in urban than in rural environments confirms that the integration of green space and contact with nature can be valuable (Figure 2.6).[35] Although it is difficult to observe altruism and its explicit relationship to design parameters, it can be shown that self-reported altruistic behaviour is more prevalent in neighbourhoods that incorporate the positive environmental and physical characteristics of space design related to diversity, proximity, accessibility and quality that have already been mentioned.[36]

Strategic opportunities for housing design

It is evident from the available research that there are no single or universal design solutions to ensure that every health parameter is optimised, and that the inhabitants and wider population will flourish. As a minimum, designers should ensure that direct physical health parameters (e.g. air quality) achieve a level that is considered 'good enough' to avoid ill health whilst not impinging on the opportunity for design to integrate wider wisdom and to support occupants in positive health behaviours.

The fact that there are numerous strategies related to different settings and users suggests that it is important to design adaptable environments. This is particularly relevant in the context of demographic change and climate change but also applies to changes in work and

Figure 2.6: 'Give': Accessible and green neighbourhoods that support the other four ways of wellbeing are more likely to encourage altruistic behaviour and increase social capital.

life styles, and the availability of new technology. Design should thus be responsive to user needs, behaviours and requirements, offering users a freedom of choice and control over their environment.

A number of rules of thumb emerge and are grouped below into key themes:

Neighbourhood and nature

There is a large amount of research related to the design of neighbourhoods that support health and wellbeing. Some of the design characteristics that emerge consistently are:

- High density mixed-use development to encourage walking and cycling ('Keep Active') to access local services ('Connect') – including access to public transport, health, social services, schools, adult education ('Keep Learning'), etc. – and reduce the reliance on the car.
- The availability of diverse public open space (in higher proportion than private gardens), including a variety of high-quality and accessible green space (for play, exercise, contemplation, allotments, socialising, etc.) and hard landscape (ideally traffic free or reduced, for play, outdoor eating, etc.). This supports all 'Five Ways' to wellbeing.
- Providing local facilities ('Keep Learning') and interest in public open space ('Take Notice') – such as a biodiverse environment (encouraging a richness of flora and fauna), seating and Wi-Fi – adds to the potential for social interaction ('Connect' and 'Give') and extends the use of the space.
- The threshold between the home and a neighbourhood, particularly in high-density scenarios, can be mediated with vegetation, both to give close contact with nature and to provide a degree of separation and privacy (Figure 2.7).
- Views of the neighbourhood and nature from the home are associated with psychological benefits and encourage social interaction ('Connect') and supervision ('Take Notice') (thus, low window sills and openable windows are valuable aspects).

Eating

Poor nutritional eating habits can lead to obesity and related health problems. The preparation and cooking of (fresh) food can become a more social activity if a kitchen is designed to enable interactions with other members of the household or community.

At a community level, the provision of neighbourhood allotments to grow fresh food is recognised as enhancing health and wellbeing due to fresh produce, physical exercise and social interaction. Furthermore, less reliance on the car for shopping and the avoidance of packaging and food miles reduce the energy and other resources required, thus improving environmental sustainability.

With respect to the design of the home, the strategy is to create a sense of 'theatre' related to cooking, and enabling 'audience participation' through the design of accessible worktops and adjacent seating (Figure 2.8). To support communal eating, and the social interactions that result, the dining area/table should be in close proximity to the kitchen. Conversely, the lounge/TV area should be less accessible from the kitchen (potentially upstairs to encourage physical exercise), limiting the temptation for 'TV-dinners' but also providing potential separation in terms of noise, odours and pollutants.

Figure 2.7: Proximity to nature is a recurring theme in research related to health, and provides a threshold between public and private space in high-density housing.

Moving and access

As we lead increasingly sedentary lifestyles, encouraging a modest level of activity becomes important in order to improve cardiac health, counteract obesity and maintain general fitness ('Keep Active'). The recommended level of activity is at least 30 minutes of moderate exercise (>3 METS [metabolic equivalents];[37] cycling or brisk walking) on five or more days per week, or 20 minutes of vigorous physical activity (>6 METS: jogging or gym exercises) three or more days per week.[38] Although gyms have become increasingly popular for some (and can also support 'Connect'), achieving improvements in fitness for all is the main goal. Moving up and down stairs is a simple and

Figure 2.8: Preparing fresh food for friends and family can become a social activity if the kitchen layout allows for this.

effective solution, which counters the tendency for choosing a bungalow for retirement (resulting in reduced exercise at a time of life when it is important to stay active, and what is colloquially referred to as 'bungalow knees'). Three-storey homes are likely to increase personal physical energy expenditure and can contribute to increased housing density, which in turn leads to other sustainable-design opportunities. Research on human energy expenditure in buildings has revealed that typical office workers are less physically active away from work, with an overall activity level marginally below the recommended levels. Thus, even modest increases in domestic and neighbourhood activity levels through design can be health-enhancing. Climbing one floor by stairs accounts for 3.3% of extra daily energy expenditure, and getting up 20 times from a seated position equates to about 10% of a healthy daily total of metabolic activity.[39] Some stealthy design strategies to 'Keep Active' are suggested:

Make circulation an enjoyable experience and provide rewards for the movement: e.g. avoid boring corridors; aim for good natural light, views, opportunities for spatial variation and encounter ('Connect'); use art, etc. (Figure 2.9). This can also create opportunities to 'Take Notice'.

The mid-20th-century German architect Hans Scharoun famously used his staircases as 'events', forming a journey through his buildings – from single dwellings to huge complex concert halls. He often angled them to deliver occupants at landings facing the direction of the next, angled, flight and had successive flights 'wind' continuously in the same direction so that people would not become disorientated as they made their way through his buildings.

Figure 2.9: Take the opportunity to encourage physical activities in and around the home.

Separate key spaces with stairs: they provide the most intense personal energy expenditure and encourage movement.

- put the living space on a different level to the kitchen/dining area
- as far as possible, avoid having WCs on every floor level (always bearing in mind that UK Building Regulations require a WC on the entrance or, alternatively, 'principal' story of any new dwelling).[40]

Conversely, for those who are physically disabled or are wheelchair users, it is clear that all housing design must accommodate this – especially in the context of an aging population. There are numerous guidance documents related to this,[41] but some key considerations include:

- accessible dimensions for circulation areas – which can contribute to a more generous experience for all
- level-access thresholds throughout – valuable, too, for families with prams
- window-sill heights to enable views out when seated – views out, especially of natural scenes, are conducive to wellbeing
- electrical sockets not too low, and worktops, handles, thermostats and light switches not too high[42] – allowing all users control over their home environment
- the potential for a lift to be installed and/or the adaptation of the home for single-floor living – e.g. bedroom and bathroom on the ground floor (also useful for temporary ill health)

Such design considerations should also incorporate strategies to ensure that partners and carers of wheelchair users are encouraged to remain active.

Indoor environmental quality

Light

Light is only briefly introduced here for completeness but will be discussed in greater technical detail later in this book. Natural light has a range of advantages over electric light, including its variability and efficiency, and creating an awareness of and link to outside conditions. Apart from being a free source of light, and thus part of an energy-efficient strategy, within a home it will animate spaces and can create drama and diversity over time. Furthermore, its benefits for physical health are now well understood and can counteract SAD. However, over-illumination can be detrimental to comfort. A number of rules of thumb emerge:

- Orient rooms used in the morning (bedrooms, bathrooms and kitchen) to the morning light to provide a dose of daylight to stimulate the circadian rhythm.
- Main habitable rooms should receive 'good' daylight (described later, in 'Sunlight penetration and shading' in chapter 5) and a key family room should have access to direct sunlight.
- Windows with high head heights provide more access to daylight via an increased sky view (which is particularly important in dense neighbourhoods) and better daylight distribution into a room.
- Bedrooms in particular should have effective blackout options to support good sleep patterns – for example, in the form of thermal shutters (for cold periods) and/or with adjustable louvres (for secure night-time ventilation in warm conditions).
- Personal control over the amount of daylight provides welcome opportunities for inhabitants to adjust conditions to suit their patterns of use, and results in a greater sense of satisfaction with their environment. Windows should offer a range of conditions – e.g. light that is from above; from the side; direct; diffuse; adaptable through shutters, louvres and blinds (Figure 2.11).

Temperature

The thermal environment is a crucial aspect of health with respect to the risk of overheating (as a result of improved insulation standards and global warming), but also regarding fuel poverty (in existing poorly maintained buildings). It is a topic that warrants close assessment, and is evaluated in detail in chapter 4.

As with light, a thermal-design strategy should create both comfortable and stimulating conditions that can exploit the climatic conditions to improve energy efficiency. The body senses the thermal environment not just in terms of air temperature but also with respect to the radiant conditions (e.g. sunlight), air movement (e.g. natural ventilation) and the conduction of heat via surface materials (i.e. wood feels warm, stone feels cool). Each of these thermal characteristics is a function of, and an opportunity for, design:

- Exploit solar radiation to create sunny places to be in on cool days, such as window seats (with warm surfaces) and sun spaces. Use heavyweight materials to absorb and retain the warmth.
- Allow the user to adapt conditions so that on hot days there are opportunities to find cool, shady places to sit on more conductive surfaces in a breeze.
- Adaptive-comfort theory (see chapter 5) reveals that thermal comfort conditions vary rather than are constant or 'optimised'. Occupant control and the adaptability of a design to suit the users'

needs and preferences as they vary over time are key factors to success.

- To cool a building down during hot spells, design openings permitting secure night-time ventilation – e.g. through louvered sections – and exploiting stack and cross ventilation principles – e.g. use the height of a staircase to enable warm air to rise and escape at the top.

Design quality

A number of other design characteristics impact on the 'Five Ways' behaviours, and these are briefly included below:

The colour of our environment: the colour of elements like interior walls can impact on our learning behaviour and thus can be used in

Figure 2.10: Daylight admitted through a range of apertures in various positions, even if quite modest in size, can result in a good distribution of light.

certain spaces to support learning. Research has suggested that 'red enhances performance on a detail-oriented task [such as doing homework], whereas blue enhances performance on a creative task [like art or social debate]'.[43]

Ceiling heights: these can play a role in our social perspective and ability to focus. Recent findings show that when people are in a low-ceilinged space they are better at focused tasks, such as studying or reading. More generous spaces prime us to feel freer and tend to lead people to engage in more abstract styles of thinking: they're better able to take a wider perspective and see what aspects are in common – particularly appropriate for social-gathering spaces.[44]

The form of space: this influences our sense of comfort and beauty. Curved forms are perceived as pleasant, and in recent experiments 'participants were more likely to judge spaces as beautiful if they were curvilinear than rectilinear'. The researchers went on to claim that this 'well-established effect of contour on aesthetic preference can be extended to architecture'.[45]

Thus blue, tall and curvilinear spaces, with views of the sky, are more likely to be pleasant, sociable and creative environments (Figure 2.12). Conversely, red, low-ceilinged and rectilinear environments are more likely to encourage focus, concentration and study.

Space standards

Our early childhood experiences have the most influence on cognitive development and subsequent health and wellbeing.[46] Designing homes with diverse settings for playing, cooking, storytelling, talking, etc. provides positive opportunities for social interaction. However, one constraint on providing diverse and stimulating conditions for social engagement in the home is a lack of internal and outdoor space.

'As every parent knows, children need space to play, develop, and do their homework. They also need privacy, especially when they are going through puberty and studying for exams. Adults need space too, to foster healthy relationships with their partners and enable them to care for their families.'[47]

Overcrowding is detrimental to positive social interactions and family relationships and is cited in surveys as the most common health theme with the strongest response.[48] It affects health and wellbeing, the spread of illness, increasing stress and the quality of relationships. Using the 'Bedroom Standard'[49] as a measure, there are more than half a million overcrowded households – affecting one child in 10 in England, with over one third of these in London.[50] There is thus a need to provide more affordable, family-sized homes that meet recommended space standards for sizes of rooms, storage and outdoor space. The size and type of affordable homes in a development, rather than simply the number of units, should match the need.

Sound

As with other aspects of environmental design, acoustic conditions (described in more detail in chapter 4) can be used to support user needs and preferences. Although noise can cause stress, acoustic contact with the neighbourhood and nature can be valuable. Similarly, within the home there are places and moments in which acoustic privacy is welcome, although complete acoustic separation is rarely required.

- To encourage 'Keep Learning' behaviours, it is important to provide quiet, calm spaces for reading and studying.
- To support activities such as music and indoor exercise without disturbing others, acoustic separation to some spaces is valuable.

Figure 2.11: Research suggests that tall, curved spaces with a view of the blue sky are associated with a more creative and sociable outlook.

- Design openable windows so that the opportunity to connect and talk with passing neighbours is possible.
- In order to exploit natural ventilation – particularly at night, in an urban environment and when quiet conditions for learning or sleeping are sought – the design should incorporate noise-attenuated air paths.
- Try to separate noise-creating items – such as washing machines and dishwashers – from living and study spaces to support social and learning activities.
- Consider the acoustics as you progress through the house: a gravel path will alert an occupant to visitors arriving; an echoey hallway and stair can signal when people are gathering; a carpeted corridor dampens the noise to the study; and soft furnishings and bedding create a tranquil environment for sleep.

Figure 2.12: A moment of delight – architecture can create opportunities and conditions for play.

Summary

Designing for wellbeing and health incorporates a plethora of opportunities and a range of criteria. The strategy proposed in this section is that designs must be 'good enough' to meet quantitative health measures but also be adaptable to, and integrated with, a broader set of principles to support health and wellbeing. There is a potential risk that in an attempt to design the technically *ideal* environment we risk reducing the importance of the stimuli that encourage occupants to be active, aware and engaged. Designs can nudge users into positive behaviours, not by making them comfortable and controlling their environment excessively closely but by providing a variety of suitable stimuli for a range of behaviours. An extreme example of this is the design for the 'Bioscleave House' by Gins and Arakawa, intended to 'strengthen life by challenging it ... to stimulate physiological and psychological renewal by creating living environments that would be intentionally uncomfortable'.[51] It achieves this aim through, amongst other things, changing floor-to-ceiling heights, distinct use of colour, textured and sloping floor surfaces, and non-standard door sizes.

One of the opportunities presented by architecture is that through the design of form, space and materiality it can order our relationships with each other and our environment by creating interactive settings for life. It can do this in such a way as to provide opportunities to improve our sense of wellbeing, enrich our lives and make them healthier and more pleasurable – for example, the shaft of sunlight in a recessed window seat that creates a moment of warmth and calm, combined with a glimpse of nature, soft and acoustically absorbent seat materials, and the tactile delight of the smooth handle that adjusts a wooden shutter. Our wellbeing is intimately linked with such moments of delight. To an extent such stimuli happen all the time, often without being recognised or designed, but when they are orchestrated throughout a building the effect is cumulative. A poor building has few such moments and leaves our lives impoverished, whereas a successful piece of architecture boasts an accumulation of many moments of delight that support health and wellbeing.

PART TWO:

Technical analysis and design synthesis

Chapter 3

The house and its surroundings

Introduction

This chapter addresses three broad themes:

1 Access to nature
2 Urban noise
3 Microclimate.

The case for access to nature

The benefits of access to natural landscape and wildlife have been widely recognised and incorporated into planning and urban design for many decades. Planning legislation such as that protecting the London Greenbelt addresses the issue on large scale. Many other regulations and codes of practice relate to smaller scales such as neighbourhood parks, provision of horticultural allotments and civic gardens (Figure 3.1).

However, until recently, there was little scientific evidence for the benefit to health and wellbeing; the motivation and support for such practices was largely based on long-held beliefs and cultural values. But in the last few decades, quantitative studies have been carried out on the influence of access to nature, in fields as diverse as the development of pre-school children and the wellbeing of dementia patients.

Most of the studies on the impact of nature on wellbeing are concerned with the proximity and the physical accessibility of open green spaces, and of varying levels of 'wildness' or, more specifically, biodiversity. In a comprehensive review by Susan Strife and Liam Downey of over 150 articles, which specifically reviewed studies on the impact of deprivation on youth development, access to nature was claimed to have physical, mental, emotional and cognitive benefits – and a positive effect on children's overall development.[1] This recent evidence gives new impetus to the introduction of access to nature into the mainstream design brief. In response we consider here the design implications in terms of the building, the site and the neighbourhood.

First however, we propose a working definition of *nature* to be that quality that *gives an impression* of being to a greater or lesser extent 'natural'. At one extreme this could be true wilderness – and at the other end of the spectrum, an urban landscape combining a minimal amount of vegetation and urban wildlife but with distant views and sky, and a lack of obviously negative properties such as pollution and dilapidation. Furthermore, we include 'cultivated nature' such as gardens and pets, which have been shown to display many of the health benefits of wild nature.

Proposed design parameters

In his book *Design for Nature in Dementia Care*, Gareth Chalfont[2] proposes a formal way of assessing nature access in care homes. Developed from Chalfont's approach we propose here a model with a systematic vocabulary that, although not fully quantified, can be used as a structured checklist to assist design.

The model illustrated in Figure 3.2 consists of ***zones***, ***links*** and ***qualities***. The total environment is divided into four zones – INSIDE, EDGE, NEAR and FAR. The zones are linked in two ways, referred to as ***Sensory*** and ***Access***.

The vocabulary is defined as follows:

INSIDE – the building interior. It can have nature content itself – e.g. indoor planting, indoor pet-keeping facilities – or even surrogate nature – e.g. photographs, paintings or sculpture.

Figure 3.1: Even heights of buildings, the availability of green and porous surfaces, and the use of light building surfaces are helpful in moderating the microclimate.

EDGE – the physical envelope and extensions of the building that link by means of doors, windows, balcony, porches or verandas. It may have nature content itself.

NEAR – the space immediately outside the building such as a patio, yard or garden. It can have nature content and structures that promote nature – e.g. planters, ponds, bird table. It will link with the INSIDE via EDGE. NEAR is usually private or semi-private.

FAR – usually off-site, large in scale, and often public – e.g. parks; farmland; wild land such as marshes, moors or mountains. It may be *contiguous* with NEAR, or *remote*.

Sensory – linked by vision, sound or smell.

- For sighted people, vision is the most important.
- It can be seen as a property of EDGE, through which the link passes.
- The links can exist from INSIDE to FAR via NEAR, or direct from INSIDE to FAR.

Access – the facility for the physical movement of the person from INSIDE to move to either NEAR or FAR.

- The link can exist between INSIDE and NEAR or INSIDE and FAR.

- If FAR is *contiguous*, then *Access* will be via EDGE and NEAR only.
- If FAR is remote, *Access* will be via EDGE and (possibly) NEAR and then an off-site area such as pathway, street, highway or even transport system (as shown in Figure 3.3), which will influence the quality of the *Access*.

Distance will also be a key parameter.

- Vertical distance is also critical for the INSIDE-to-NEAR link.
- It is far less likely that a person will have good use of a garden if they are living in a fourth-floor rather than a ground-floor flat.
- It might be better to put resources into a shared patio garden at fourth-floor level, or even individual balconies.

QUALITY can be ascribed to all four environments and the links. In the case of the FAR environment this will include scenic variety and biodiversity (the two being closely related). Clearly an area of rolling hills, outcrops of woodland and hedgerows is of greater restorative value, both to view remotely and to walk amongst, than modern 'industrial' farmland. The NEAR environment is more dependent on architectural elements and the way that the immediate surroundings to the building attract wildlife and horticultural activity. For the INSIDE environment in domestic spaces the nature content is limited in architectural terms and relates more to occupant decisions about furnishing and contents. However, in multi-residential blocks there may be opportunities for indoor planting, water features and nature-related art in spaces such as foyers, staircases and other communal areas. Many studies on indoor planting in non-residential buildings indicate that it has significantly beneficial effects.

QUALITY is just as important when applied to links. *Sensory* INSIDE / NEAR and INSIDE / FAR mainly applies to windows. The importance of positioning windows in relation to views and to seating areas is well known to architects, but is often neglected in response to demands for standardising layouts or even structural constraints. Special consideration should be given to supervision of children from within the house. A garden or patio that can be supervised from inside (possibly from a kitchen) by a parent will be much more useful than one where the parent has to enter the garden area in order to check up on the children. An openable window will facilitate audible communication, as well as ventilation.

However, when considering fenestration in relation to children, openable windows can create the serious hazard risk of falling out – even ground-floor windows sometimes being sufficiently high up to cause serious injury. Where there is the risk of child access, or indeed risk to anyone due to low sill height, opening-limiters should be incorporated. This can also allow intruder security to be maintained whilst open.

QUALITY applied to the *Access* link is very important. It may be associated with the EDGE – i.e. the design and location of the door or doors into the NEAR environment, relative to the occupancy of the rooms. For example, good access would be achieved where the main living area is linked to the garden or patio by doors also providing a visual link. This is not uncommon, often involving wide openings with sliding patio doors.

The QUALITY of the *Access* link to the FAR environment is equally important. If the FAR and NEAR environments are contiguous the link will be short – i.e. a gateway in the boundary of the NEAR environment. This is highly desirable but probably a luxury that few urban dwellers enjoy. Far more frequently, the link will involve a journey through a built-up area of no nature value and generally used by the public. It may also involve

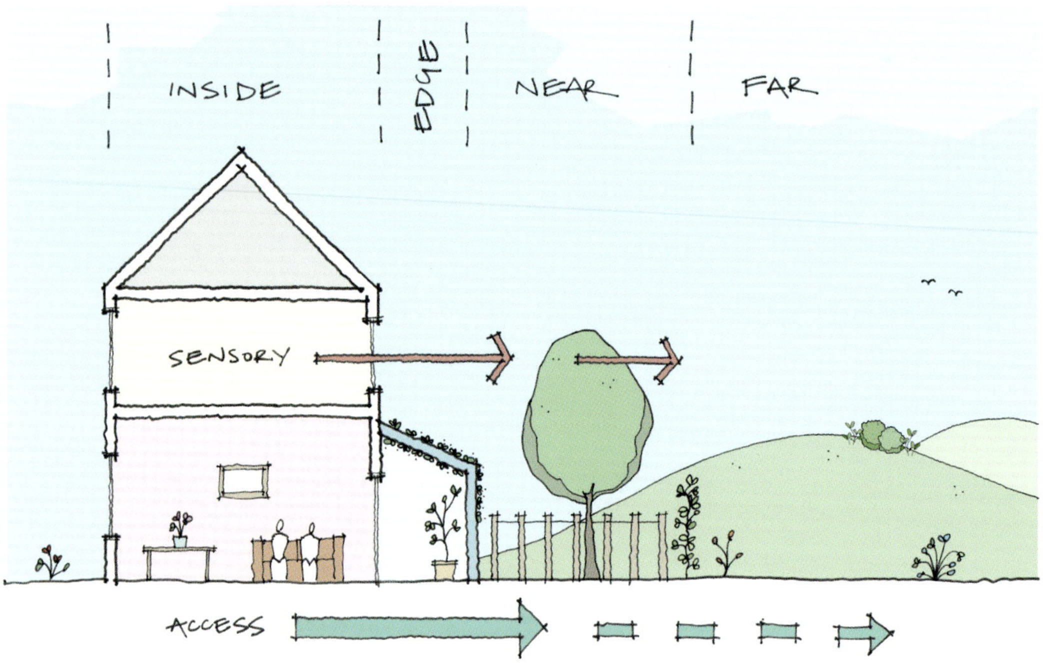

Figure 3.2: Section illustrating the configuration of zones and links for a low-rise residence with contiguous NEAR *and* FAR *zone.*

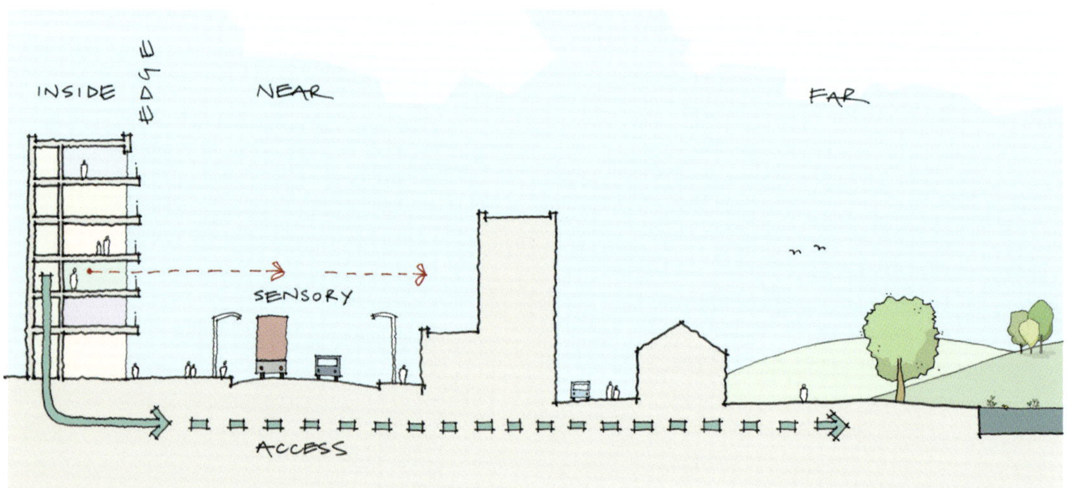

Figure 3.3: Section illustrating the configuration of zones and links for a high-rise flat showing balcony as EDGE*, the* NEAR *zone with no nature value, and a remote* FAR *zone with high nature value. The* Sensory *link to nature is compromised by the noise of the street and the obstruction of the buildings opposite. The* Access *link is degraded by requiring a lift, crossing a busy road and possibly an 'unsafe' neighbourhood.*

hazards from road traffic and perceived or real dangers to children and vulnerable people from anti-social and criminal behaviour. Or it may simply be unpleasant due to ugly surroundings, noise and atmospheric pollution.

Another critical factor is distance. In the Strategy for Improved Nature Access for London (Greater London Authority 2008), the target maximum distance is set at 1km. This equates to about 10–15 minutes' walking.[3] Clearly, distance and environmental quality interact – a longer walk in pleasant urban surroundings could be regarded as being equivalent to a short unpleasant one. However, for vulnerable people – e.g. children, or the old – a single feature such as a dangerous road crossing, or a steep hill, might be critical.

Access for wildlife and gardening

This refers to access for wildlife *to* the NEAR zone. Sensory contact (usually visual) with active wildlife – birds, mammals, insects and plants – is greatly valued. Most birds have access by air, and even the common urban species such as pigeons provide a valuable nature resource in otherwise dire urban environments.

There is already much design guidance available (e.g. from English Nature[4]) on provision for birds to nest and feed on or close to buildings. Indeed, much of the loss of common garden species is blamed on changes of building design that deny birds roosting and nesting places, and changes in garden style that replace flowerbeds planted

Figure 3.4: An example of a view of nature in both the INSIDE *and* FAR *ranges.*

Figure 3.5: Even in a relatively confined urban context, balconies provide opportunities for green space that are a boon for wildlife and people.

with flowering annuals and shrubs with tarmac hard standing for a car, and paved patios with gas heaters.

Whilst designers cannot guarantee that occupants will become avid gardeners, there is much that can be done to encourage this – via the provision of planting opportunities, even in confined spaces such as balconies, and surfaces with easy access where birds can be fed or pots can be placed safely. In multi-storey buildings where plant pots or containers may be positioned near balcony edges or on window sills, provision must be made to prevent accidental falls of pots or containers. Gardening activity will also have an impact on insect population, which in turn will attract birds. Whilst many insects may be of limited interest to the average occupant, and some regarded as pests, there is almost universal appreciation of butterflies, moths and dragonflies.

Gardening as an activity is itself of benefit to wellbeing. One of the areas in which this has been demonstrated is in the care of vulnerable people such as the elderly and the mentally ill, where it is referred to as Social and Therapeutic Horticulture (STH)[5]. With an ageing population, provision of gardening facilities (however small) in all housing must be a high priority. There is also much evidence that children benefit from practical involvement with plants.

In public areas of high nature value, interpretation boards are found to have a positive effect on nature appreciation. This principle could be brought down to a more local level in multi-occupancy buildings, providing information on wildlife likely to be seen from the building.

Pet keeping

Pet keeping is widespread in Europe. In 2012 the European petfood industry federation claimed that about 25% of households have a dog or cat. Including fish, reptiles, birds and small mammals adds approximately another 25% of households. This statistic is fairly uniform across European states.

There seems to be near-universal agreement that pets are of therapeutic value (Figure 3.6). A study by Deborah Wells has shown that many human disorders – including behavioural, emotional and physical conditions – have improved with regular contact with pets, particularly dogs or cats.[6] There are many cases in which animals (typically dogs) provide actual physical assistance; the best-known example are guide dogs for blind people. Increasingly dogs are being used for hearing, detecting medical conditions and general assistance. Whilst their main role may thus be utilitarian, there is no doubt that these animals frequently also fulfil an emotional role for their disadvantaged owners.

Pet owning may be of particular value in urban situations, where density and lifestyle create stress. However, these same locations make pet keeping more difficult. Provisions that facilitate pet keeping often overlap with nature-promoting design of the NEAR and FAR environments. The provision of a secure outdoor space and access to it may be a critical factor in cat or dog ownership, and the provision of space for a pond, rabbit cages or aviary similarly important. Dog walking, an almost essential part of dog ownership, has needs in terms of access to FAR nature environments, but also has great benefits since it in itself promotes health through exercise as well as exposing the walker to nature and ensuring that the nature provision is well used.

Figure 3.6: Pet keeping has been proven to provide therapeutic value.

Pet keeping can create a number of local inter-neighbour problems. These usually centre around noise, hygiene and perceived danger. To some extent these are design issues – some of which, such as noise, are already covered by codes of practice and in 'Urban noise and its mitigation' later in this chapter. Some are generated by anti-social and thoughtless behaviour, however, and cannot be designed out.

These might seem highly specific issues outside the normal architectural design sphere. However, bearing in mind the prevalence of pet keeping, and its benefits to health and wellbeing, it seems appropriate to bring some design effort to bear on this neglected area.

Key points

- Access to nature should be considered systematically over a wide range of scales, from the building interior to the neighbourhood.
- Sensory links should be considered as well as actual physical access.
- Activities such as gardening that can provide a substitute for wild nature, and can attract wild flora and fauna, should be facilitated in or close to the building.
- The above activities are particularly important for occupants with limited independence – such as children, and the elderly and vulnerable.
- Pet keeping should also be seen as a common and justified part of living, and provided for wherever possible.

Urban noise and its mitigation

Sound climate and noise

The ambient sound climate of a site can range from birdsong and the distant mooing of cows to the perpetual roar of a motorway. The emotional response of a person living there may be equally diverse, from delight to serious stress and sleep disturbance. Thus, it is essential in appraising a site that the sound climate is assessed or – in some cases, where major changes are expected – predicted.

Noise is defined as unwanted sound. This implies a value judgement by the hearer. In the examples above, we would expect natural sounds to be received positively by most people whilst most would regard traffic noise negatively. Thus, at the urban scale, we will concentrate on commonly regarded noise sources including traffic, aircraft and industrial noise. Noise from neighbouring occupants – speech, music and general domestic activity – can also cause irritation and stress, and will be dealt with later.

Road-traffic noise

Much work has been carried out on traffic noise in urban areas and reliable methods of predicting noise levels in relation to the nature of the road traffic (vehicles per hour, type of vehicles) and distance from the road.[7] Figure 3.7 shows the noise-level contours emitted from an unscreened compared with a screened road.

The decibel (dB) is a unit of sound-energy intensity on a logarithmic scale in order to cover a very large range of intensities. The 'A' in dBA denotes that the sound has been weighted over its frequency range, roughly according to the sensitivity of the human ear. Most sounds that concern us are not one-off events, so we need a way of measuring sound over a period of time. Accordingly, L_{10} is the

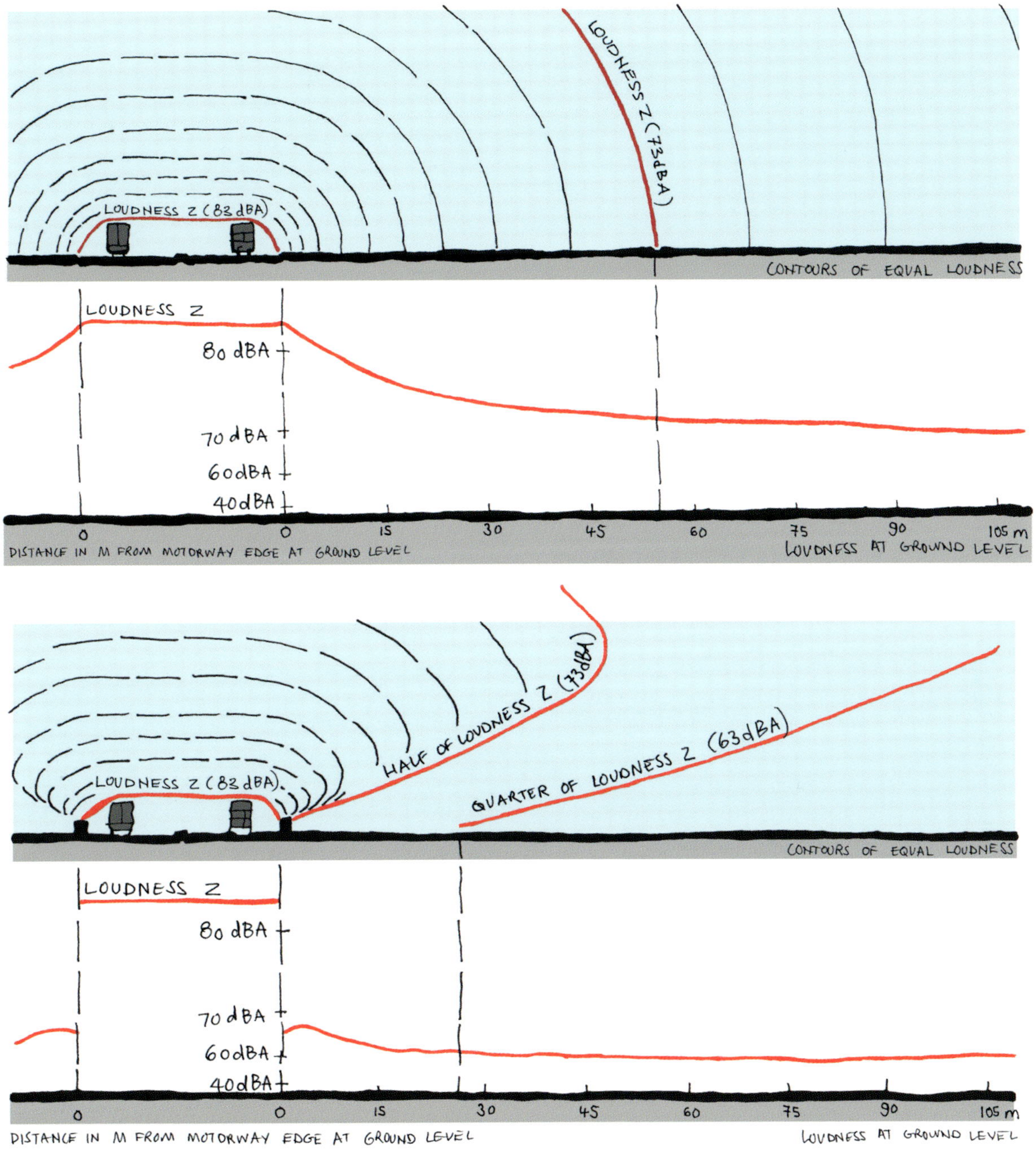

Figure 3.7: Noise-level profiles (at ground level) across a motorway WITHOUT noise barrier (top) and WITH noise barrier (bottom).

sound level exceeded for 10% of the time, and L_{90} that exceeded for 90% of the time. They can be considered to represent 'average peak intensity' and 'average background level' respectively.

An important rule is that for every doubling of distance, the noise from continuously flowing traffic reduces by 3 dBA. Thus, if the level at 10m from an urban highway is L_{10} 72 dBA, in order to drop to L_{10} 54 dBA, as recommended for residential development, the distance would have to be doubled six times – a distance of 640 metres! Clearly this rules out distance as a strategy for traffic-noise control.

Acoustic barriers

Figure 3.7 shows the effect of an acoustic barrier close to the source. Another solution is to provide a barrier close to the receiver, but this is the less favoured option since the former protects a larger area and, potentially, many receivers. However, the close-to-source option may not always be possible due to established urban fabric – e.g. shops, commercial buildings and the gaps between them. Strategies to deal with this are shown in Figure 3.11.

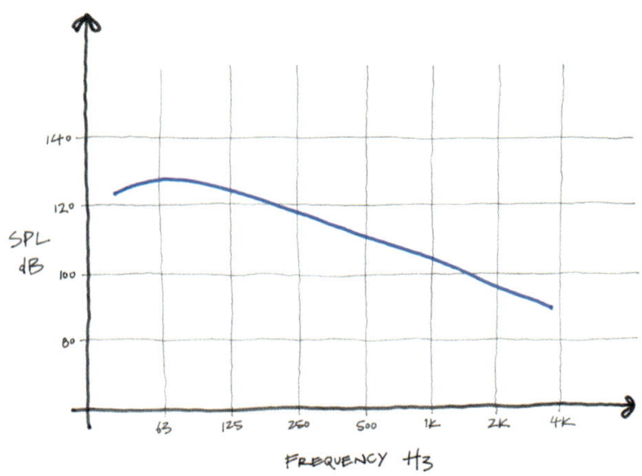

Figure 3.8: Frequency spectrum of road-traffic noise. 'SPL' on the vertical axis refers to Sound Pressure Level.

Problems could arise when the carriageway is elevated, due to 'line of sight' interaction over large distances. However, close shielding is now more likely to be built into the road structure. It must also be noted that 'line of sight' screening only works for high frequencies. Fortunately, this is where the most annoying part of the sound is located on the spectrum (see Figure 3.8); however, low frequency noise, which can diffract around barriers, may still cause annoyance.

Not all sound barriers perform equally effectively. The best barriers are rigid and heavyweight – for example, earth banks and concrete walls. Rigid lightweight structures will screen high frequencies but not low frequencies. Lightweight porous structures, such as vegetation, are very poor at screening any sound. This is summarised in Figure 3.9.

Other noise sources

There are similarities and subtle differences between railways and roads as noise sources. One difference is that trains do not flow at the same density as road traffic; rather, their noise is manifest as a series of discrete events as individual trains pass. The maximum noise level occurs when the train is at its closest, which may in some existing cases be very close. When the distance is smaller than the length of the train, then for the time that the train is passing it behaves as a linear source, like the motorway, and the 3dBA-per-doubling-of-distance rule applies. But if the receiver is much further away than the length of the train, it behaves as a 'point source' and the sound reduces at a rate of 6dBA per doubling of distance.

An isolated static noise source, such as a factory building or a building site, also acts as a point source (except when very close) and so the 6dBA rule applies as well.

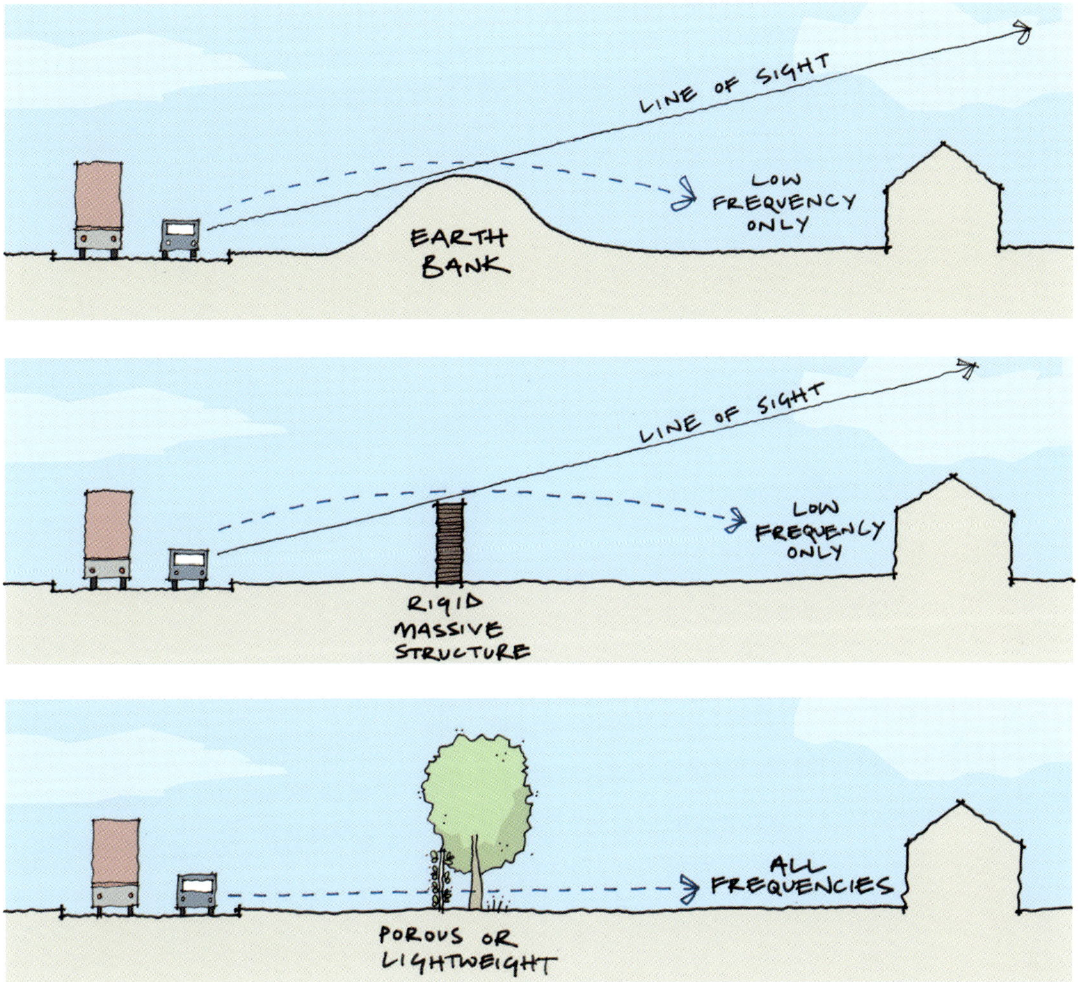

Figure 3.9: Performance of various noise screens.

In all cases, the screening principles are similar to those applying to road traffic. But there may be further complications due to reflections, as discussed below and shown in Figure 3.11. If in a project there is any chance of noise being a problem, the recommendation is to carry out a survey or obtain existing noise-survey data.

Aircraft noise

Aircraft noise, usually associated with taking off and landing at an airport located close to a large conurbation, by its very nature affects a large population. Whilst restrictions on night flying may reduce disturbance, it will by no means be eradicated.[8]

Aircraft noise is also present as discrete events. For areas near airfields, maps are usually available with L_{eq} contours. L_{eq} is used to specify intermittent noises such as aircraft noise, and is defined as the sound intensity of a steady sound that has the same level of annoyance as the

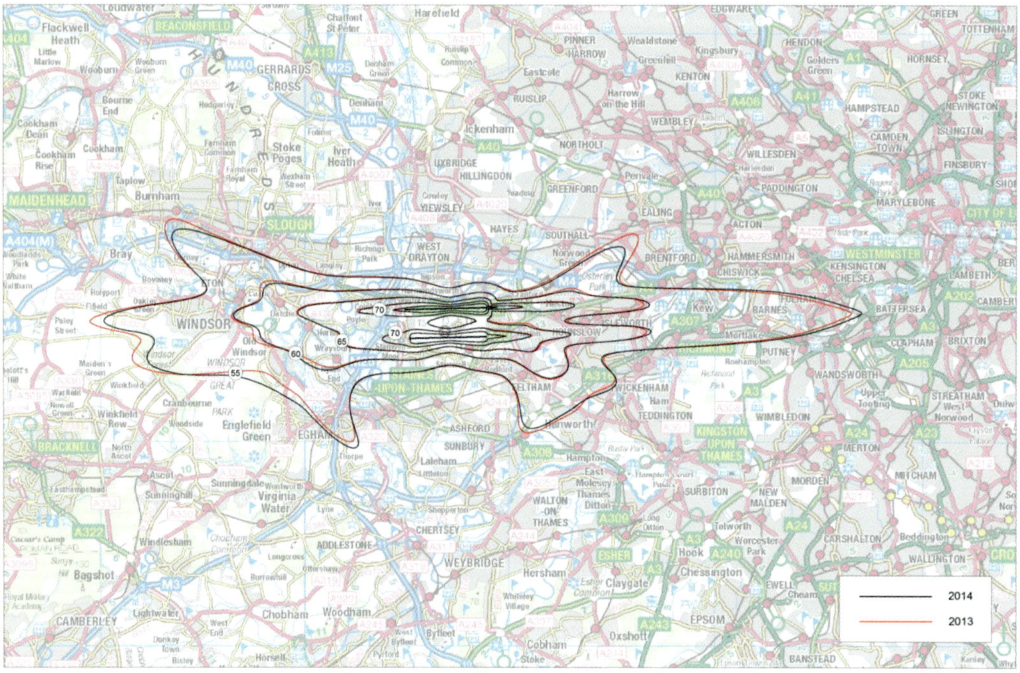

Figure 3.10: L_{eq} contours for Heathrow Airport.[9]

intermittent sound under consideration. A typical L_{eq} map is shown in Figure 3.10.

The loudest noise is generated as an aircraft takes off and climbs. Thus, if these conditions are catered for, noise associated with landing and taxiing will be covered. During the take-off and climb, the sound source moves from being on the ground, and therefore affecting vertical façades, to being overhead (or partially overhead), when roofs will be the main sound-reducing elements. And whilst screening strategies can be effective in the former case, it will generally be the roof alone that provides sound reduction for buildings under the flight path. Many more residences beneath the flight path will be affected by overhead noise than those adjacent to the airfield affected by lateral sound.

The detailed design of a building's envelope, and window and door openings, to achieve sufficient sound reduction will be described in chapter 4.

Screening noise by site layout

Housing is probably the most sensitive building type to noise, so it may be possible to screen a residential area using other buildings less sensitive to noise. Another possibility is self-screening, whereby a principal fenestrated façade and outdoor space face away from the noise source. The noise-facing façade may still have small areas of sealed glazing, but it is recommended that acoustic expertise be sought since the noise-reduction performance of the façade is sensitive to detailed design.

A further consideration is reflection. Bear in mind that even if the noise-facing façade has good noise-reduction performance, this could be undermined by reflected sound entering the 'open' façade, as shown in Figure 3.11.

Figure 3.11: *(a) Noise screening by a building,*
(b) self-screening,
(c) the problem of reflection.
Note: a useful rule is that medium- and high-frequency sounds (equivalent to middle C and upwards) will be reflected from hard surfaces larger than $1\ m^2$, like light from a mirror.

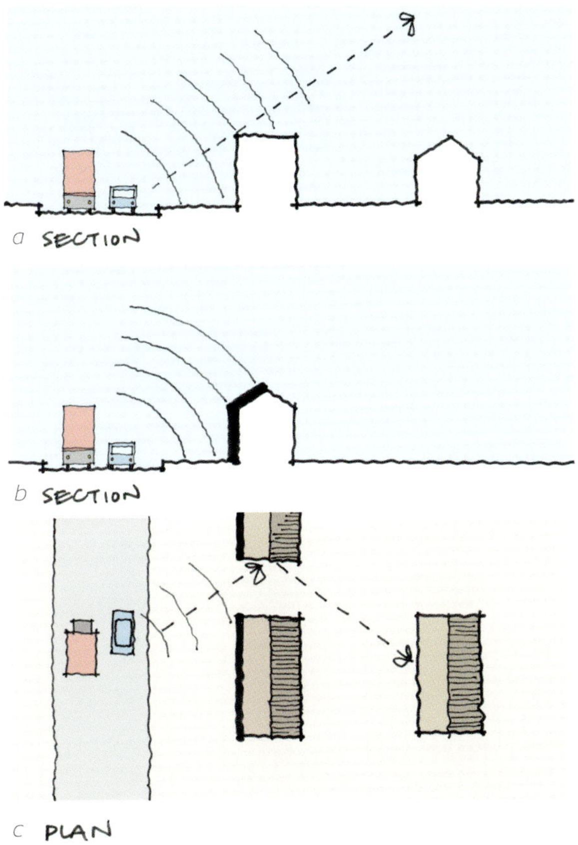

Noise problems at the neighbourhood scale

Inter-neighbour conflict is frequently caused by noise – speech, music, activities such as gardening or DIY – all acceptable to the originator of the noise, but often a source of great stress to the neighbour.[10] Some of these issues are also dealt with in 'Noise and its contrtol by the building envelope and structure'. (see chapter 4).

Problems often arise where common circulation spaces, such as passageways and staircases, pass close to openable windows or ventilator openings. The sound of people passing a window as an intermittent and relatively rare event is generally not a problem. However, it is important to avoid invitations to congregate in spaces to which private space has acoustic links.

There is a potential conflict here, since public areas may also be green areas that provide valuable views from private space. This may be partially resolved by window design as discussed in 'Fenestration and view' in chapter 4.

The classic noisy-neighbour problem is often found in older properties where party walls separating two dwellings are of inadequate acoustic performance. For new buildings, UK Building Regulations require a minimum Sound Reduction Index (SRI) of 53 dBA for a party wall. This means that raised-voice noise of 80 dBA would be transmitted to the adjoining property at 27 dBA and, as this would be 8 dBA below the recommended background of 35 dBA for bedrooms, it is unlikely to cause annoyance. Fortunately, for sound transmitted through massive structures such as party walls it is the high frequencies that are attenuated the most. This means that speech is filtered to a low unintelligible murmuring, which is usually found less annoying irrespective of the actual level.

However, this raises the interesting point that although it may meet criteria for average conditions there will be occasions where the instantaneous background noise level drops below that of the transmitted sound and, once identified, the transmitted sound may cause annoyance. It demonstrates the important point that lowering background noise is not necessarily a good thing since it also lowers the threshold for identifiable sounds that may be irritating – not because of their loudness but because of the information that they carry.

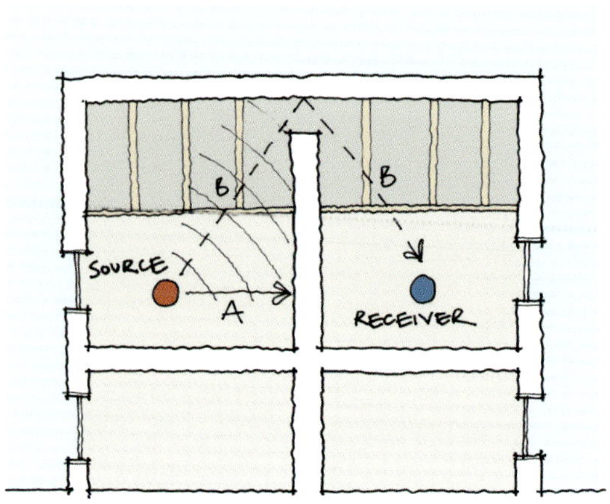

Figure 3.12: Sound-transmission routes between adjacent houses:
(A) direct transmission through party wall,
(B) flanking sound via roof space

Due to the non-linear nature of the perception of sound by the ear, sound transmitted through cracks, gaps and fissures, or small areas of much lower SRI, can be significant. For example, a 10 m^2 party wall of a room with an SRI of 48 dBA can have this reduced to 30 dBA by cracks with a total area of only 20 cm^2, which if located behind a skirting board or under a timber floor may well pass unnoticed. Furthermore, this sound will contain higher frequencies. This issue is discussed in more detail in 'Effect of gaps on the sound reduction of a partition' in chapter 4.

This underlines the importance of ensuring the integrity of any separating structure, and avoiding flanking and/or reflective sound routes. This principle is illustrated in Figure 3.12. It is especially relevant to retrofit projects where it is vital that the correct diagnosis is made before resources are committed to a solution.

The issue of the nature, or information content, of sound is an important factor in assessing the stress caused by noise. 'Impersonal' noises such as road traffic, rail or industrial noise generally cause far less annoyance for a given level than neighbour noises such as conversation or entertainment (such as music), or household activity such as vacuum cleaning. A further complication is that an individual's response to a neighbour's noise is dependent on the social relationship between them.

Key points

- Sites must be assessed for existing and future noise developments. Urban noise sources include traffic, rail, industry and aircraft. If the noise climate is indicated to fail appropriate criteria, mitigating actions must be taken to meet them or surpass them.
- Distance is a poor mitigating strategy. Screening close to the source (if the source is well defined) is the best; if not possible, screening close to the receiving site should be employed.
- The performance of screening is dependent on the frequency of the sound and the nature (shape and materials) of the screen. Screening can be undermined by reflections and the bending (diffraction) of low frequencies around the screen.
- Where screening is not possible, the building envelope must be designed to give the required sound reduction to meet criteria for the interior.
- Avoid creating outdoor noise sources such as refuse-collection areas, car parking and congregation points close to private space with low acoustic isolation – e.g. windows.
- For noise nuisance from adjacent properties, it is essential to identify the acoustic path (transmission or flanking) and to identify any defects (such as cracks) in the separating structure.

The microclimate

Long-established responses and newer hazards

The climatic parameters that effect human comfort, health and wellbeing can be broadly classified under two categories:

1. those that humans have a long-established response to, such as temperature, solar radiation, precipitation and humidity
2. those that are essentially a product of modern urban activity, such as gaseous and particulate pollution, road traffic and aircraft noise.

Adaptive comfort-seeking behaviour will normally ensure that exposure to the first-category parameter does not reach levels where permanent damage results. Thus, people will walk on the shady side of a street, will seek shelter from the rain and snow, and even escape to the countryside when urban conditions become unbearable. But exceptions to this happy state of affairs will be all too common in groups of people whose economic situation or social constraints prevent them from taking appropriate action. Thus, to some for whom an urban heatwave means more iced coffees, visits to the swimming pool and lunches in the park, it will be bearable; for others, housebound in top-floor poorly insulated flats, it may prove fatal. Global warming has only added to this concern in that, under its effects, heatwaves are ever-more likely to occur.

The second category of environmental parameter is even more insidious. Because we have not evolved to detect the danger of, for example, particulate carbon from diesel engines, or nitrogen oxides and VOCs from internal-combustion engines, we may be unaware of their presence and take no evasive action. Even when we are conscious of the agent, such as aircraft or traffic noise, we may not be aware of the long-term effect of sleep disturbance that results. In other words, although we have a normal comfort response economic and social pressures my result in our ignoring climatic parameters. Such thwarted adaptive action is a form of chronic stress that will cause long-term harm.

The means of mitigating the impact of climatic stress are dealt with in more detail in chapter 5. Here, we briefly review just two urban climatic phenomena that may have significant health implications for a proportion of the population:

- the heat-island effect
- urban wind patterns

For access to a large body of research on urban climate, the reader is referred to the review paper by Moonen, et al.[11]

The heat-island effect

The term 'heat-island effect' is used to describe the temperature increase observed within a city when compared with the outlying rural area. Typical values range from 1.1 K (degree Kelvin) to 6.5 K, listed by the United Nations Intergovernmental Panel on Climate Change (IPPC),[12] with extreme values as high as 14 K claimed by Santamouris, et al.[13] However, these are not average effects but extreme differences. One characteristic of the urban temperature cycle is that due to the thermal inertia of massive buildings it warms slowly, reaching a peak in the afternoon and evening (the latter impinging on sleep comfort during early night time).

The effect is present year-round, and in winter can be seen as an advantage – increasing outdoor comfort and decreasing the demand for heating. However, due in part to global warming and the ever-increasing dependence on air conditioning, most concern is with the summer case.

The main causes of the summer heat island are:

- increased net solar-radiation absorption
- reduced evaporation
- reduced convection
- increased thermal storage
- anthropogenic heat production.

Increased solar absorption

This is due mainly to the roughness of the urban surface. Solar radiation (visible and infrared – IR) gets trapped in the cavities between buildings by multiple reflections – it is a common experience that a rough surface looks darker than a smooth surface of the same colour. The reflectance of building surfaces (albedo) will also influence their net absorption; however, even cities that may superficially look light in colour, seen from the solar angle, have quite a low mean reflectance.[14]

Reduced evaporation

Except in desert regions, most rural areas are vegetated; transpiration of water cools leaves, which create a cooling effect in the air. Even where soil is not vegetated, groundwater is free to evaporate from exposed soil surfaces. This contrasts with the urban surface where typically there is a smaller fraction of vegetated area, and most hard surfaces are relatively impervious to water vapour.[15] Furthermore, whereas rain is free to soak into the rural surface most urban water is drained away to join an underground watercourse.

Reduced convection

Urban surface roughness reduces wind speeds at street level, which results in lower convective losses. This, together with the higher thermal admittance (the combination of high thermal capacity and conductivity) of building materials compared with that of vegetation, results in urban night-time temperatures in particular remaining significantly higher than rural equivalents.

Increased thermal storage

These higher night-time temperatures may have a direct impact on sleep: it is harder to take adaptive action at night. Thus, this effect has special significance for residential buildings. Conventional bedding means that at least half of the body is well insulated. Ambient noise and/or pollution may create problems impeding adequate natural ventilation. This means that it is all the more important to reduce daytime solar gains into the thermal mass of a building by shading, as described in more detail in 'Sunlight penetration and shading' in chapter 5.

Anthropogenic heat production

This comes from building heating and cooling, lighting, transport and industrial processes. There seems to be general agreement in the literature that the contribution of anthropomorphic waste heat is significant in winter in high latitudes, where the solar contribution is very weak.[16] On the other hand, it is small compared with solar effects during summer.

In one very specific type of case, heat output from air-conditioning units themselves could be significant: this occurs in high-rise buildings where condenser fan coils positioned on the perimeter wall, on several floors, can generate a plume of warm air that will subsequently affect floors above.

Urban wind patterns

Urban wind flow has been investigated by field studies and with scale models in wind tunnels and CFD (Computational Fluid Dynamics) modelling,[17] mainly to assess its impact on the ventilation of pollutants from streets, as shown in Figure 3.13. Other issues include outdoor thermal

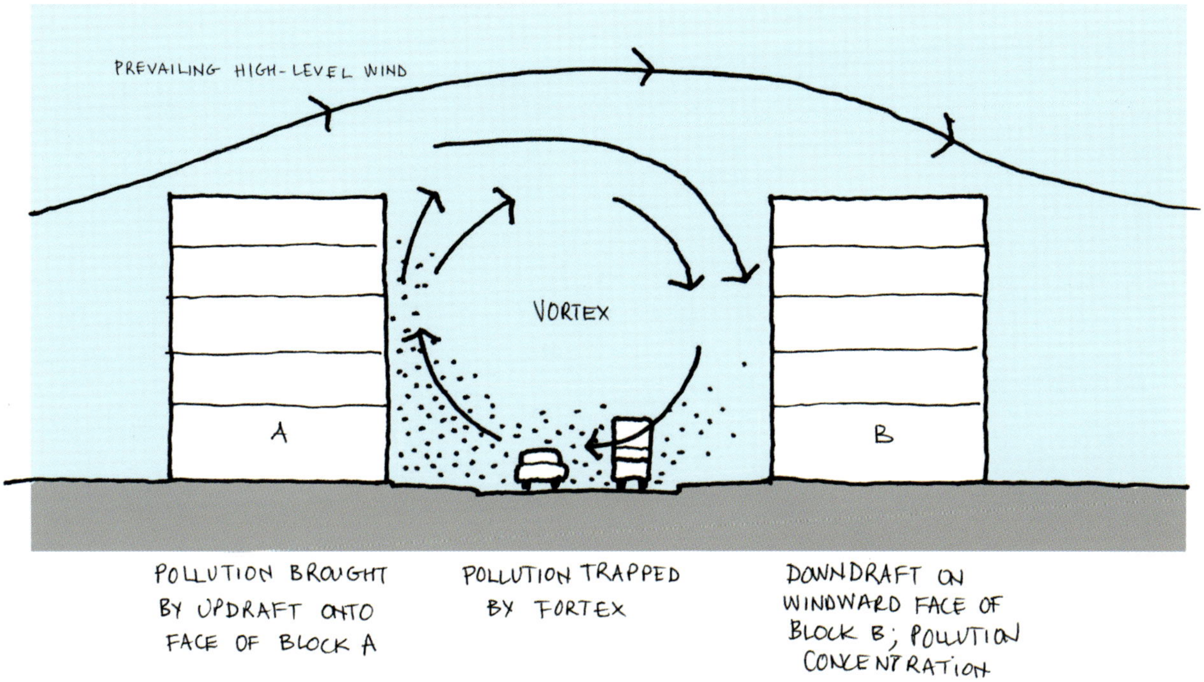

Figure 3.13: Characteristic urban wind-flow patterns and their interaction with street pollution from traffic.

comfort (wind chill in winter and cooling in summer), and the presence of high-speed gusts that might impede pedestrians.

Urban pollution is without doubt a key area of concern. The combination of carbon monoxide, nitrogen dioxide, ozone, volatile hydrocarbons and particulates – all products from the internal-combustion engine associated with transport – has serious impacts on health. This issue is being addressed by many city administrations, partly by limiting traffic; by demanding improved emission standards from vehicles; and, in some cases, allowing only electric vehicles. Much measured data is available for major cities, and computer models can be used to predict levels from inputs including traffic flow, prevailing wind and street geometry. Typical model output maps[18] are shown in Figure 3.14.

Wind can affect street comfort directly. Unlike porous vegetation, which reduces wind speed, solid bluff objects such as buildings deflect and redirect it. In doing so, unstable flow paths are set up that can create gusts of wind speeds greater than that of the prevailing wind. If these occur at the populated street level, this can cause discomfort and distress to pedestrians.

These conditions are often produced when there are constricted gaps or passageways between or through buildings, or where a tall slab-sided building protrudes into the faster-moving air above.

Design responses

In most cases, the designer of small-scale residential developments will have few opportunities to influence existing conditions. In which case, it is essential that they become aware of the local microclimate and make appropriate responses as described in chapters 4 and 5. However, it would be a pity if the design itself contributed to current urban-climate problems, and in this respect the designer should follow this guidance (Figure 3.1):

- Use light colour (high-reflectance) finishes to hard ground surfaces, walls and roofs.
- Provide as much green or porous surface as possible.
- Avoid excessively tall slab blocks protruding above the average building height.

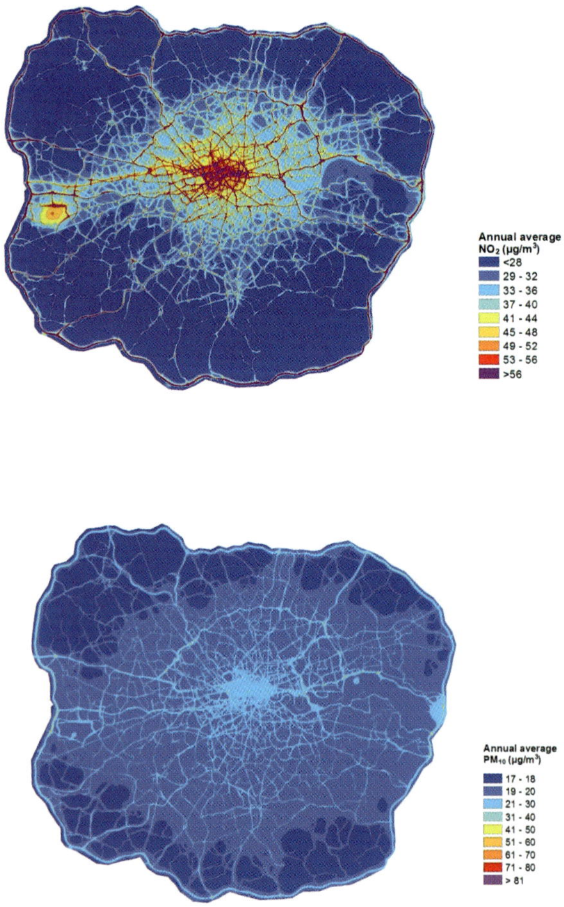

Figure 3.14: Annual mean nitrogen dioxide (NO_2) and particulates (PM_{10}) concentrations for Greater London produced by ADMS-Urban modelling software.

Key points

- The heat-island effect can elevate peak temperatures by as much as 6 °C, and in particular can increase night-time temperature – interfering with sleep in residential buildings.
- Winter minimum temperatures will also be higher by 2 °C–5 °C owing to the effect.
- Average wind speeds will generally be much lower at ground level but may be more turbulent. For buildings rising above the average building height, wind speeds will increase considerably.
- Pollution from road traffic in streets can be highly variable and relate to local wind patterns.

Chapter 4

The building envelope

Introduction

This chapter examines in detail the following aspects of the external building envelope:

1 Materials
2 Fenestration
3 Thermal performance
4 Noise

Materials

Environmental impact and embodied energy/carbon of building materials

Choice of building materials can have an impact on the environment at both local and global scales. Local effects can include the appearance of a building; its performance structurally, thermally and acoustically; and its resistance to weather and how it ages (Figure 4.1). It can also have an impact on the internal environment – in particular, air quality, which is dealt with in more detail in chapter 5.

The global impact of materials used in building has attracted intense interest. The commonest indicator of impact is their embodied energy, or embodied carbon. These are related through the burning of fossil carbon fuel (and its subsequent emission of CO_2) to generate heat in order to, for example, make steel, cement or glass; or to generate electricity for the production of aluminium. Plastics and resins, derived from oil, also involve large amounts of energy in their production.

On the other hand, some materials need much less energy in their production and processing. Wood is a notable case, in that the material itself not only needs no fossil carbon in its production but its growth also takes CO_2 out of the atmosphere. Other plant-based fibres have the same qualities. However, wood and timber products still need energy for processing and transport, and this can amount to a significant contribution to embodied energy. Natural mineral materials such as stone and clay may not need energy in their production but still involve large amounts of energy in their mining or extraction, and transport. A slate floor may look natural and 'environment friendly', but if the slate has been shipped to the UK from China it probably is not.

Table 4.1 shows examples of embodied carbon for some common building material per kilogram and per cubic metre. The choice of structural materials, and finishes, can have a major impact on embodied carbon. But it is more complicated than indicated by the table, due to the wide range of quantities of different materials that constitute a building, and their wide range of

Material	CO_2/kg	CO_2/m³
Aluminium	*8.24*	*22.25*
Steel	*1.40*	*10.55*
Bricks	*0.23*	*0.39*
Reinforced concrete	*0.18*	*0.46*
Expanded polystyrene	*3.45*	*0.10*
GRP (glass-reinforced plastic)	*8.10*	*10.10*
Plywood	*0.42*	*0.17*
Softwood	*0.19*	*0.067*

Table 4.1: Embodied energy and carbon for common building materials.[1]

Figure 4.1: The choice of building materials is rich and varied: it impacts on appearance, interior environmental conditions and the global environment.

densities. Thus, it is better to compare equivalent *structural units* as in this example.

'... a wooden floor beam requires 80 megajoules (MJ) of energy per square metre of floor space and emits 4 kg CO_2 during its production. By comparison, a square metre of floor space supported by a steel beam requires 516 MJ and emits 40 kg of CO_2, and a concrete slab floor requires 290 MJ and emits 27 kg of CO_2.[2]

One important consideration is the use of a high-energy material to reduce the energy consumption of a building; a good example is the case of thermal-insulation materials, and the concept of a 'payback period' is useful here. This is highly relevant in refurbishment projects, in which payback time can be used to prioritise different measures. Figure 4.2 shows a graph of the total carbon emissions (embodied plus in-use) for three situations:

1 an existing building
2 the same building with insulation and conservation measures
3 a new 'green' building.

It illustrates the point that the payback time, which may or may not be acceptable, should influence the retrofit strategy. Retrofitting conservation to the existing building pays back in 9 years (A), a new building compared with existing takes 16 years (B), but a new building compared with the retrofit case takes nearly 43 years (C).

There are other kinds of environmental impact apart from CO_2 emission and its effect on global warming. Other atmospheric pollutants resulting from industrial processes are of concern – such as sulphur dioxide (SO_2), which causes acid rain, and nitrogen oxides (NO_x), which have direct impacts on respiratory health.

Several databases that assess environmental impacts of building materials have been assembled to inform the designer of the impact of material choices. For example, the Embodied Carbon Database was commissioned by WRAP (the Waste and Resources Action Programme) and the UK Green Building Council to capture embodied carbon data for whole buildings.[3] The aim is to provide a free and publicly available resource for building professionals to be able to benchmark their designs to a far greater extent, and to provide more detailed comparative data than was previously available to any single company or individual.

Visual qualities of materials and their interpretation

This is a very subjective area, with all the dangers of adopting a prescriptive theory. There is no doubt that certain materials have strong

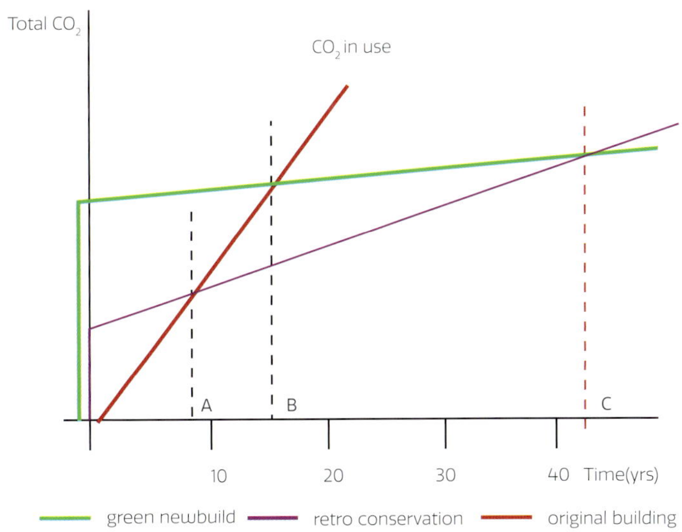

Figure 4.2: Total carbon emissions (embodied and in-use) for an existing uninsulated building, the same building with retrofit conservation and a new 'green' super-insulated building, showing CO_2 'payback' time.

associations with health, both physical and psychological. However, the actual relationship may be false or, at least, context sensitive. For example, the use of natural materials such as wood and stone in a domestic kitchen, where food is prepared for family and friends, may be seen quite differently from its use in a fast-food outlet, where hi-tech materials such as laminates and stainless steel may be much more appreciated.

The explanation may be that in the former case levels of hygiene are in the control of the occupants and users of the kitchen, whereas in the second case the consumers of the food have no control, and (perhaps erroneously) feel more secure with materials that are associated with high levels of sterility.

Moving outside the building, vernacular, Modernist and hi-tech architectural approaches all have their supporters. It is generally accepted that pride in the appearance of our dwellings is a positive response, and it follows that the provision of good designs that engender this make a contribution to wellbeing. However, it is difficult to associate this with any particular material choice other than that which is covered by generally good, appropriate design. It is interesting that the eco-friendly healthy building *look* is just as strongly associated with hi-tech materials as traditional or vernacular ones.

Figure 4.3: A hi-tech look is strongly associated with a healthy building, as is a building that uses more traditional materials lilke timber and stone (see Figure 4.1)

An important property of a material is how its appearance changes under the influence of weather. It is well recognized that some aesthetic schemes show their age poorly, relying on their pristine condition to express their quality. Other materials 'grow old gracefully' and may even improve with age. And it is not just the property of a single material that demands consideration but rather that of an ensemble of materials. For example, the stain on a timber or stone wall caused by the run-off from a metal roof may or may not be regarded as elegant; in most cases, probably not.

The relevance here is that if the absence of restorative maintenance gives the visual impression of a run-down neglected building, this will probably have a negative impact on the wellbeing of the occupant and possibly be a source of friction with neighbours. It is essential then that the designer is aware of how appearance will change, what maintenance is necessary if the changes are unacceptable and that these are made known to the occupant. It also follows that where there is a risk that sufficient maintenance is unlikely, special consideration should be given to materials and finishes that require little or none.

Impact on local environment

Historically, materials such as asbestos and lead created serious health problems, mainly for people manufacturing building materials and components, and installing them. Since their banning, attention has switched to modern materials that remain exposed to the interiors of buildings, which, due to emissions of volatile compounds or particulates, are likely to influence the health of the occupants. These include paints, adhesives, polymer fibres, plastic laminates and even cleaning materials.

This issue is dealt with in more detail in 'Pollution sources – occupants, contents, activities, outdoor air' in chapter 5.

In the event of a building fire, emissions from burning materials can be highly toxic and play a major role in the injury and death of occupants and firefighters. This is a highly specialised topic and the reader is referred to specialist literature, some of which will be found in the 'Further Reading' section of this book.

Key points

- Embodied carbon is a good indicator of global environmental impact, and should influence choice of construction and materials.
- Embodied carbon should also be considered when deciding between refurbishment or new-build.
- The visual quality of materials can convey mood and atmosphere to interiors and exteriors, which may influence occupants' sense of wellbeing.
- How materials age, visually as well as functionally, and what the implications are for maintenance should be taken into account when designing and specifying.

Fenestration and orientation

Definition of orientation

This term is often used rather loosely, but in the context of fenestration we will define it as the direction that the largest area of glazing in a building faces. This is because the impact of solar radiation on an interior is dominated by a glazed façade, as is the availability of daylight. Furthermore, especially with openable windows, noise and pollution penetration are much more likely through a glazed façade. Thus, 'orientation' in a strictly geometric sense – i.e. in relation to the major and minor axes of the building – will not be used.

Sunlight and daylight access

It is commonly understood that east-facing rooms receive morning sun; south-facing spaces, midday sun; and west-facing ones, evening sun – and this has traditionally had some influence on the choice of function for rooms. However, there are several complications that make such simple rules difficult to apply.

Sunlight in rooms

Firstly, it is important to establish whether sunlight penetration is a positive or negative attribute. The technical advantage of passive solar gain as a form of space heating is long established. But if this is merely displacing well-controlled mechanical heating, it cannot in itself be relevant to occupants' health and wellbeing since the heating effect will not be separable from the mechanical heating. Secondly, in warmer seasons and climates solar gain can be a disadvantage, causing discomfort due to overheating. Visual discomfort owing to glare is also an issue. Both overheating and glare can be overcome by well-designed shading devices, as described in more detail in 'Sunlight penetration and shading' in chapter 5. Here, we are cautioning against over-glazed façades simply on the grounds of maximising solar gain.

However, the presence of sunlight in rooms for psychological benefit is equally well established, particularly in cooler climates. And more specifically, its role in establishing circadian rhythm, vitamin D production and even its biocidal effects have been recognised – especially in health buildings such as hospitals. This is discussed in more detail in 'The Case for Daylight' in chapter 5.

In principle, sunlight will penetrate windows whenever the solar azimuth angle is equal to the window normal angle, plus or minus 90°. However, the transmission of glass varies with angle of incidence (*i*) as shown in Figure 4.4 and, furthermore, obstructions such as frames, glazing bars and reveals mean that the glass surface is shaded for larger angles of incidence. In practical terms, solar gain will be much reduced at *i* greater than about 45°, but visually and psychologically the impact of sunlight penetration may still be present in the room.

In medium to high latitudes, sunrise azimuth angle and time vary significantly with season – for example, at 55°N sunrise is approximately in the southeast in winter at 09.00 hrs and in the northeast in summer at 04.30 hrs. Thus, an east-facing bedroom that is appropriately sunlit from about 07.00 hrs in spring and autumn could be sunlit from 04.30 in midsummer unless some geometric or movable controls are provided. In winter, where at this latitude the maximum solar elevation is only about 15° at midday, only south-

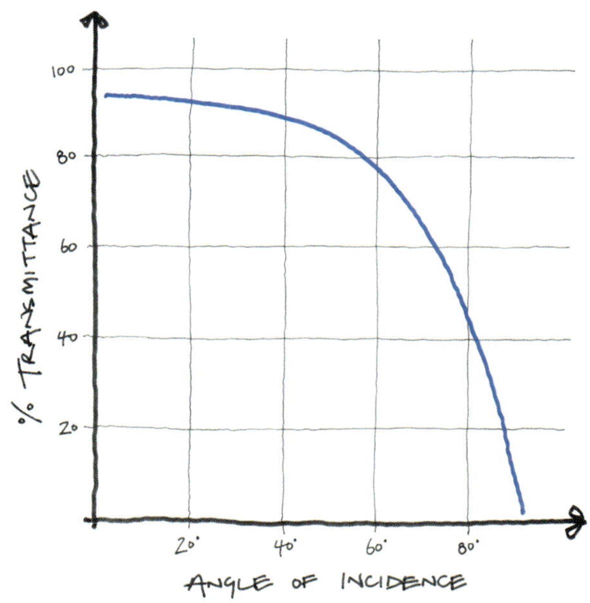

Figure 4.4: Transmission of direct sunlight through glazing as function of angle of incidence, i (but note that there will be a further reduction of transmission due to the shadowing effect of glazing bars, mullions and the depth of any external window reveals).

facing rooms will receive significant amounts of sunlight penetration.

Even where the programming of room use in relation to sunlight penetration is a clear objective, a further complication is that at urban densities, street layouts and obstructions will make it highly unlikely that the optimum criteria can be met in more than a small proportion of a group of housing units.

Thus, in many cases it may be best not to let orientation considerations dominate the massing and geometry of a building, but rather to accept the status quo and then ensure that detailed design maximises opportunities to make some kind of contact with sunlight. This could be by providing visual access to sunlit surfaces and the use of high external reflectances as an alternative to sunlight penetration, as in Figure 4.5. This practice can go a long way to bringing the psychological and, to some extent, the objective benefits of sunlight to occupants. In some cases, a choice between a view and a *solar orientation* may have to be made – in which case, the quality of the view must be balanced against the benefits of sunlight.

Daylight

The provision of good daylighting in rooms is a high priority in housing, from both health-and-wellbeing and energy-conservation viewpoints. The design of interiors to achieve this is dealt with in chapter 5. Here, we consider the provision of daylight on a building scale – i.e. the exposure of a building to the luminous sky.

Usually considered to be the light from the diffuse sky, 'daylight' strictly includes light from the sun's disc and the so-called circumsolar area where forward-scattered sunlight increase the luminosity of the sky compared with the rest of the sky dome. The CIE[4] overcast sky (a mathematically defined model of sky luminance for standard design independent of latitude) is symmetrical about the vertical axis, and thus has no sensitivity to orientation. In contrast to this, a clear sunny sky is highly asymmetrical; at midday a southern façade (in the northern hemisphere) receives up to seven times more intensity than a northern façade. *Standard skies* include a range of sky distributions according to the climate and weather, and even pollution level. The standard *average sky* is the condition in which a weighted time average of the different types is taken. These sky types may be offered by daylight-modelling software for use with detailed daylight design, which can be further extended to provide dynamic analysis often referred to as *climate based*.

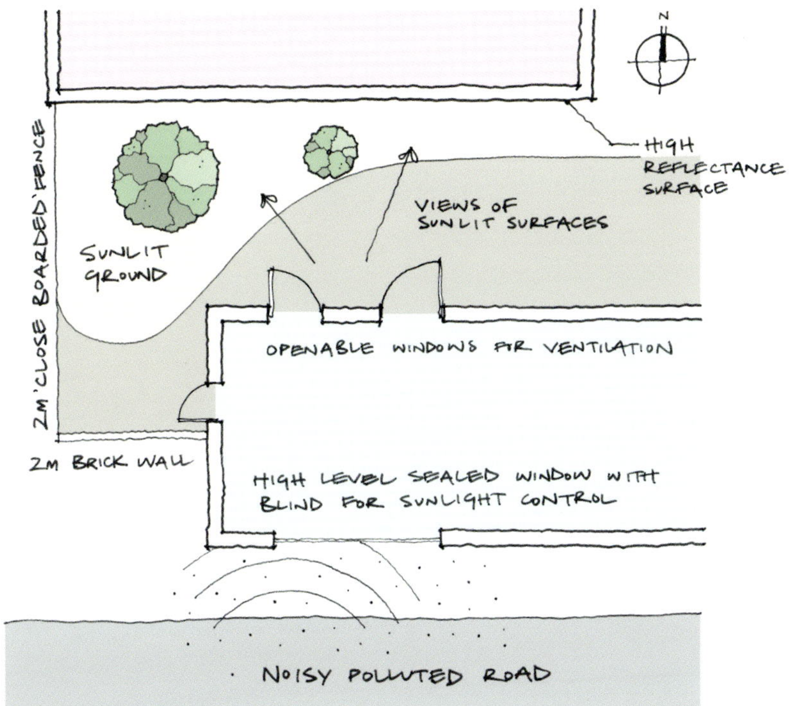

Figure 4.5: Sunlight-access strategy with unfavourable site constraints.

Apart from passive heating from solar gain in winter, daylight illumination at a level higher than that needed for visual tasks has an important role in circadian-rhythm synchronisation and the avoidance of SAD. This will be discussed in 'The Case for Daylight' in chapter 5.

Orientation (the azimuth angle) does not have a very strong influence on the availability of the minimum useful daylight for visual tasks other than, in some cases, the avoidance of direct sunlight by choosing a northern orientation for rooms such as studios and workshops. A useful rule of thumb at initial planning stage, for the UK (derived from the work of D.R.G. Hunt),[5] is that the availability of useful daylight on vertical surfaces is roughly in the following proportions: North 0.77, East 1.04, South 1.20 and West 1.00 – i.e. a south-facing façade has 20% more useful daylight than a west-facing façade.

However, the elevation of the visible sky patch above the horizon is more important since, as described by the CIE sky, the sky luminance is on average three times greater overhead than it is at the horizon. This results in a rooflight typically capturing two to three times more light than the same area of vertical glazing. It also follows that glazing in an upper wall surface is far more effective than glazing closer to the floor.

Sky View Factor

The availability of daylight to a building envelope is dependent upon the amount of the sky visible from the relevant point on the building. This can be calculated provided the geometry of the surrounding obstructions (usually other buildings) can be defined, and is known as the Sky View Factor (SVF). Working with a digital elevation model of an area of townscape, either existing or proposed, it is relatively easy to compute the SVF for any point on the surface of a subject building or the ground. An example is shown in Figure 4.6. Note that SVF does not take account of the non-uniform illuminance of real skies.

Such modelling can give a good impression of how an occupant might respond when standing at a window, in terms of their perception of urban density, and thus could influence their sense of wellbeing.

However, it is not the sole indicator of daylight conditions inside the building, as this is also influenced by the angle of incidence to the glass

LEICESTER WATERFRONT DEVELOPMENT:
Sky View Factors used to test compliance with daylighting regulations in Sustainable Homes directive

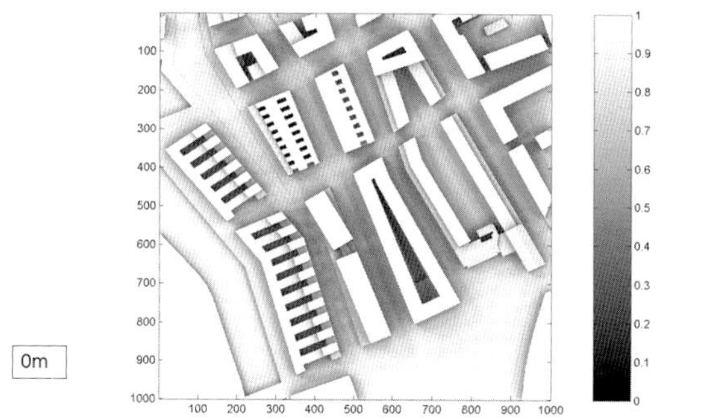

Figure a

LEICESTER WATERFRONT DEVELOPMENT:
Sky View Factors used to test compliance with daylighting regulations in Sustainable Homes directive

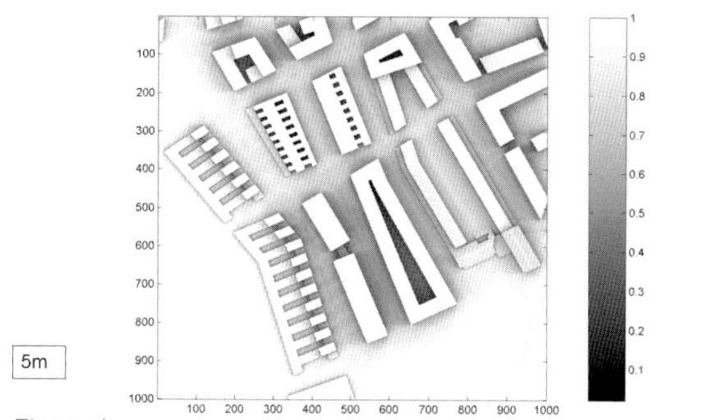

Figure b

Figure 4.6: Sky View Factor (SVF) map of proposed housing development in greyscale at ground level and 5 metres. SVF mapped from digital elevation data. By calculating Daylight Factor (DF – see chapter 5), glazing can be positioned and dimensioned to meet DF targets.

and reflected light from strongly illuminated external surfaces. This has already been discussed in the context of direct sunlight.

Noise exposure

Values of Sound Reduction Index (SRI) in Table 4.2 show that in most cases the glazed part of an envelope is the most transmissive to sound. Thus, the orientation of the glazed façade to any noise sources has a large impact. The mitigation of this by screening and sound reduction by the envelope has already been dealt with in 'Urban noise and its mitigation' in chapter 3.

Construction	*125 Hz*	*500 Hz*	*2,000 Hz*
Single brick wall	*32*	*40*	*55*
Double brick wall with ties	*37*	*47*	*62*
100 mm studs 25 mm exterior plaster 12 mm plasterboard	*21*	*41*	*47*
Hollow-core door, weather sealed	*19*	*25*	*20*
Single 4 mm glass	*18*	*26*	*33*
Double 4 mm glass, 6 mm airspace	*21*	*24*	*44*

Table 4.2: Sound Reduction Index (SRI) for construction elements for various frequencies. Note that even when closed, openable windows may have a lower SRI value than glass. In combination (e.g. a window in a masonry wall) the overall SRI is dominated by the smaller SRI (e.g. the window).

Here we draw attention to the fact that the detailed design of windows has considerable bearing on their SRI. Furthermore, without exception their SRI is reduced significantly when they are open. This has implications for ventilation strategies in relation to urban noise sources such as roads and railways, and regional sources such as airports. Where glazed façades face significant noise sources, occupants will be much less likely to us them for ventilation. This can lead to problems of overheating and poor air quality.

The problem can be mitigated to some extent by the design of window openings that both screen and absorb sound without completely obstructing airflow, as in Figure 4.7. These will inevitably increase costs, and will have practical limitations on airflow efficiency for a given area.

However, as with the case of reconciling the relative values of solar access and view, compromises may have to be made. Fine distant views of nature may be available from a façade facing a noisy urban highway. Even with sound-attenuating openable windows, indoor noise levels could be unacceptable. But if natural ventilation could be obtained by openings in another quieter façade, then the main windows could be sealed for high sound reduction.

Natural ventilation

Ventilation

Ventilation has three main functions:

1. to maintain air quality
2. to remove heat from a room
3. to remove heat from a person by direct air movement.

These functions will be discussed in more detail in 'Ventilation and indoor air quality' in chapter 5. Ventilation can be achieved by natural or mechanical means. Natural ventilation is driven

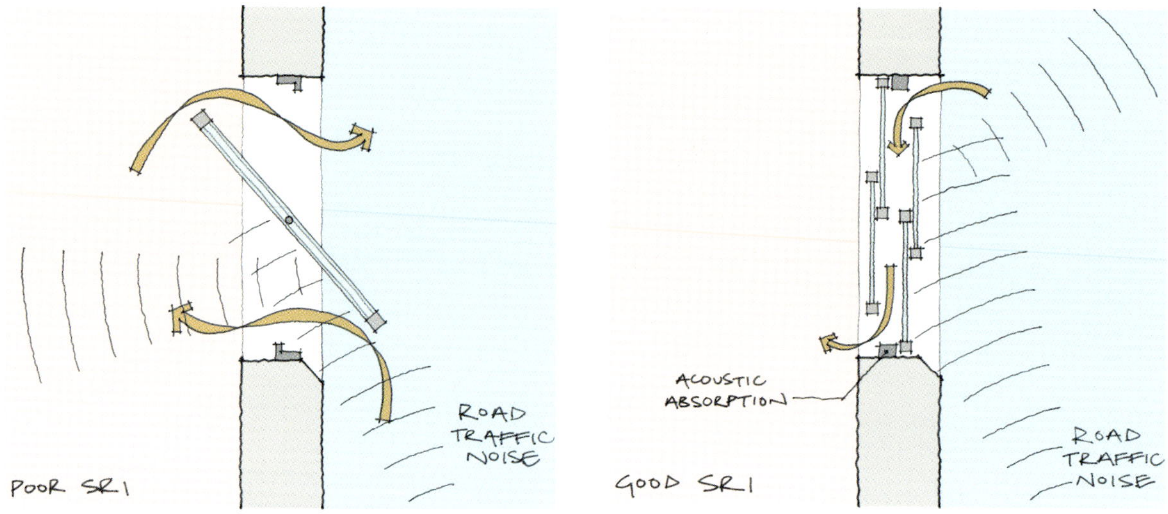

Figure 4.7: Poor (left) and good (right) window design in relation to Sound Reduction Index and ventilation.

by two separate mechanisms: thermal buoyancy (sometimes referred to as the stack effect), and wind pressure (see Figures 5.12 and 5.13 in the next chapter).

In reality, particularly in urban environments, the wind-pressure mechanism is complex (more complex than Figure 5.15 in chapter 5 might suggest). Surrounding buildings interact with the airflow, causing turbulence that results in the wind strength and direction varying continuously (Figure 4.8). Furthermore, even on open sites the undisturbed wind direction varies over different timescales.

Another factor is that the driving forces, wind strength and temperature difference between inside and outside are highly variable and will have no correlation with ventilation demand. For example, on a cold windy day the demand for ventilation is likely to be far less than on a still warm day. The result of this is that fixed ventilation openings positioned and sized for simplistic assumptions on wind directions and temperatures are unlikely to be successful.

A useful physics principle to apply is that air will flow from a point of higher pressure to a point at lower pressure. Thus, a flow of air will be set up between any two openings that are at differing pressures, even though they may both be negative or positive relative to the ambient air. This generates the rule that where actual pressure distributions are hard to predict, it is better to provide many small openings distributed over several façades of a building than a few large openings. And when considering the large variability of the driving forces, it follows that the openings should be controllable.

One further consideration is that air will only flow into and out of two openings in a building envelope if there is a relatively low-resistance internal pathway between them. Clearly, partitions and doors will interrupt a flow path, and in avoiding this there may be compromises around privacy, noise and pollution control.

Infiltration

Traditional buildings were very leaky, with cracks around door and window frames, eaves junctions, services connections and constructional joints. Furthermore, many older houses used open fires or stoves, which greatly increased the extraction of air and, therefore, its replacement with fresh air. By modern standards these buildings were over-ventilated, with average air-change rates[6] as high as 3 ac/h (air changes per hour), and this usually meant that, in the winter at least, little or no further (intentional) ventilation was required.

The resulting heat loss by infiltration has now been identified as a major target for conservation measures, and modern and refurbished houses will have infiltration rates typically less than 0.75 ac/h and often lower. The ventilation functions for air quality and thermal comfort now have to be provided by special elements, ranging from openable windows to sophisticated pressure-controlled vents. The latter are able to compensate for the variability of the wind and thermally induced pressure.

However, this places a new burden on the architect. Designing ventilation, particularly natural ventilation, is now part of the brief and it has to be considered in the overall spatial design of the building – not just a services 'add-on'. In the past, except in a few specialist buildings, it was largely left to chance.

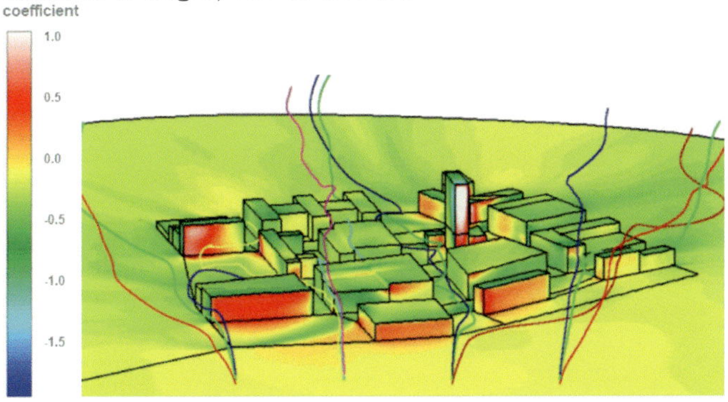

Figure 4.8: False-colour map of wind-induced pressure on building surfaces in a dense urban setting. This pattern will change minute by minute due to turbulence.

The issue is summed up by the maxim first coined by the Building Research Establishment: 'build tight, ventilate right'. Here, *ventilate right* refers to the matching of the location and time of the ventilation to the need. This is dealt with in chapter 5.

Key points

- Openings for natural ventilation should be multiple and well distributed over the envelope, and on different façades if possible.
- Openings should be easily controllable.
- The use of natural-ventilation openings may be compromised if they open onto noise or other pollution sources.

Fenestration and view

The strong case for providing views of nature, and of interesting townscape and sky, has been made in 'The case for access to nature' and the potential conflict between views and mutual privacy with neighbours in 'Urban noise and its mitigation' (both in chapter 3). Here, we only address the issue of conflict between other orientating factors such as solar access and noise.

When considering the orientation of a major block, a decision may have to be made whether to optimise view or solar access. Solar access is seen by some designers to be an essential part of sustainable design, and given high priority. However, in our view, if the optimum solar orientation turns its back on high-quality views this is too high a price to pay.

We have to bear in mind that with recent developments in energy technology, passive solar heating is not as essential as it was a few decades ago. In a modern highly insulated house or apartment, with carefully controlled ventilation

and perhaps heat recovery, the contribution of passive solar heating in absolute terms is quite small. Furthermore, a strongly solar response (large areas of south-facing glazing) will bring its own problems of overheating and make demands for large areas of movable shading.

Nevertheless, as discussed above in 'Sunlight and daylight access', it is recognised that solar penetration has value other than that of its energy. Where housing units have multiple façades, it may be possible to separate out the functions – e.g. providing a main northwest-facing window for view and ventilation, whilst a smaller high-level window facing southwest lets in evening sunshine.

There is evidence that the most effective window is one that combines all functions.[7] The acts of seeing through, then opening, then feeling the outside air, then hearing the outside sounds and maybe sensing outdoor smells have a value that exceeds the sum of their parts. There is also evidence to suggest that when people make an environmental decision for themselves, they are prepared to make large compromises. For example, taking a seat at a pavement café table would expose a person to noise and pollution conditions that they would never accept inside a building.

This suggests that even when the main view orientation is also a noise source, well-designed openable windows that have good sound-reduction performance when closed, should be provided together with separate sound-attenuating vents. This strategy is particularly appropriate to bedrooms. It provides the choice to the occupant of enjoying the full multiple function of the openable window, albeit with noise, and also gives the alternative of separating out the functions. It was the basis of a policy in London in the 1970s when new urban highways were cut through hitherto quiet residential areas.

Figure 4.9: Even in non-ideal situations where noise, privacy or orientation can be challenging, it can still be possible to design with building form and fenestration to provide access to light, fresh air and greenery.

One further factor generating a need for a specific view is the supervision of children. It has already been pointed out in 'The case for access to nature' in chapter 3 that a garden or patio will be much better used as a play area if it can be supervised, visually at least, from within the house.

An openable window will allow better verbal communication, even if it is not normally used for ventilation due to noise conditions. Consideration also has to be made for the function of the room from which the visual connection is made. Ideally this would be a room such as a kitchen or living area, which would be occupied during the day when young children would most likely be playing outside.

Louvred blinds can easily be set to provide shade from direct sun and privacy, but still allow views (in this case, usually below a horizontal angle) to the outside.

Key points

- Windows have several functions: ingress of diffuse daylight, provision of near and distant views, ventilation if openable and solar gain.
- Do not let any one function dominate decisions on window orientation without weighing up the benefits and dis-benefits.
- There may be technical and architectural ways of separating out window functions.
- Giving the occupant choices, lessens negative impacts.

Thermal performance and energy balance

In cool climates, heated buildings lose heat to the environment primarily by two mechanisms:

1 conduction through the envelope
2 ventilation.

Heat loss by conduction

Conduction is influenced by the choice and thickness of materials in the envelope. It is well known that structural materials such as brick are poor insulators whilst lightweight materials such as mineral wool and cellular plastics are highly insulating. In modern practice, this has led to the separating out of the two functions – structure and insulation – and results in envelope conduction values more than 10 times less than those in uninsulated buildings of 50 years ago.

For buildings in warm climates, where it is desirable for the interior to be cooler than the exterior (either by passive or mechanical means), heat flow is from outside to inside – and similar benefits to those in cooler climes are gained by high insulation values. However, due to the effect of solar radiation on the surface of the building, which creates large but transient temperature differences that are sensitive to building orientation and geometry, the spatial distribution of the insulation and the reflectance of the building's surfaces, and shading, becomes important. For example, adding insulation in the roof will have much more effect than the same amount of insulation added to walls that are shaded by deep overhanging eaves.

Figure 4.10: Another innovative design that brings light and air to a non-ideal setting.

Another difference is that whilst internal gains in winter in cooler climates can be useful to offset heat demand, in summer they may lead to overheating – and this can be exacerbated by high levels of insulation. However, if high ventilation rates can be provided, this is not a valid case against high insulation values. The basic calculations relating to heat flow (heat losses and solar gains) is given in 'Heat loss by ventilation'.

How much insulation?

Calculations show that the value of extra insulation diminishes as more is applied. Figure 4.11 shows the effect on the U-value of successively adding a further 75 mm layer of insulation. It also shows that the incremental saving due to each successive layer reduces drastically, in the proportions 84% to 7% to 2%.

This suggests that 'more is not always the better solution', and that an 'optimum' thickness should quite reasonably be influenced by other considerations such as insulation cost, the structure needed to contain it and the space it takes up. This is particularly the case when retrofitting insulation to existing buildings.

Heat loss by ventilation

In a cool climate, the provision of fresh air to replace indoor air creates heat loss simply by the need to warm up the incoming air. As insulation standards have increased, so the focus has shifted to reducing this heat loss by reducing over-ventilation; this was discussed in 'Natural ventilation' earlier in this chapter..

A second strategy is to adopt heat recovery. This is a mechanical process that extracts heat directly from the outgoing vitiated air and uses it to preheat the incoming fresh air. The method usually necessitates the partial mechanical handling of air – i.e. fans and ducts.

It is an essential part of the design strategies for PassivHaus.[8]

The same mechanism of ventilation heat loss that is curbed in heated buildings is of benefit in rejecting unwanted heat. This unwanted heat may be due to solar or occupancy gains, and ventilation becomes an important strategy for maintaining thermal comfort. This is dealt with in 'Ventilation and indoor air quality' in chapter 5.

Because heat losses by conduction and ventilation are both proportional to the temperature difference between inside and outside, they are often combined to give a single heat-loss coefficient. This can be used with monthly temperature data to calculate the monthly or annual heating energy required.

Heat loss can be calculated from the following equation:

$$W_t = (\Sigma\, U.A + 0.33\ .\ N.V).\Delta t$$

Where W_t is the total heat flow in watts, U and A are the conductance (or U-value) and area of the individual elements respectively, Σ indicates that these are summed, N is the number of air changes per hour and V is the volume of the room or building.

Any heat gain to the room will cause a rise in temperature, and thus create a temperature difference, Δt, between inside and outside. This will cause heat to flow through the envelope to the outside; Δt will rise until the heat loss through the envelope exactly equals the heat gain – this condition is known as *heat balance*.

It can be seen that when the total energy input is equal to the heat loss, the temperature difference – and, hence, the room temperature – can be controlled either by the heat input W_t or the ventilation rate N. In practical terms, this might mean pulling down the shades to keep the sun

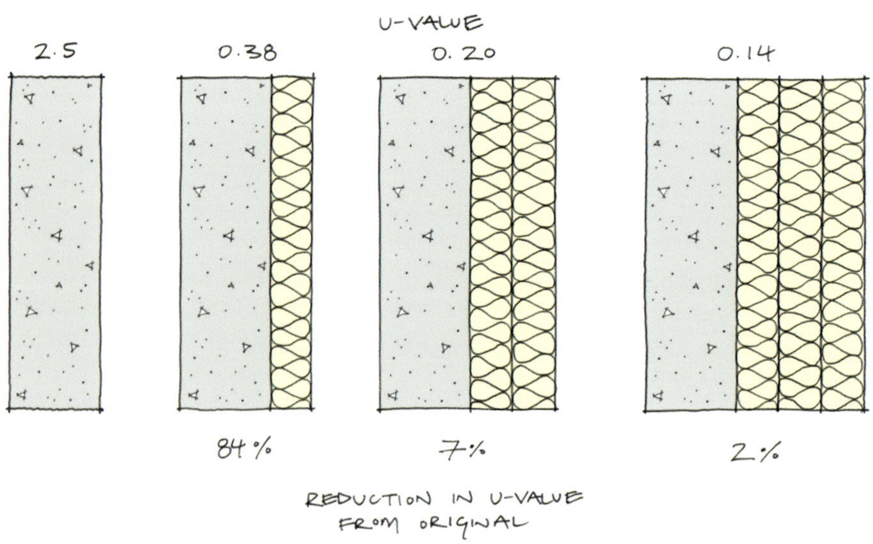

Figure 4.11: Effect of successive 75 mm layers of insulation to an uninsulated wall.

out or opening the window to increase the ventilation rate.

Solar gain W_s can be calculated from the equation:

$$W_s = I.\ A.sgf$$

> Where I is the incident radiation in W/m^2; A is the area of glazing; and sgf is the solar gain factor, which is mainly influenced by the transmittance of the window.

Note that I is for the actual orientation of the window and is dependent on the orientation of the window relative to the direction of the sunbeam – viz: $I = I_b.\ Cos\ \theta$, where I_b is the intensity in the direction of the beam, and θ is the angle between the normal to the surface and the beam. This means that sun falling on surfaces at a very oblique angle will cause much less solar gain than when it is at right angles. This is further reduced by the glass and the shading effect of glazing bars, as was described in 'Sunlight and daylight access' earlier in this chapter.

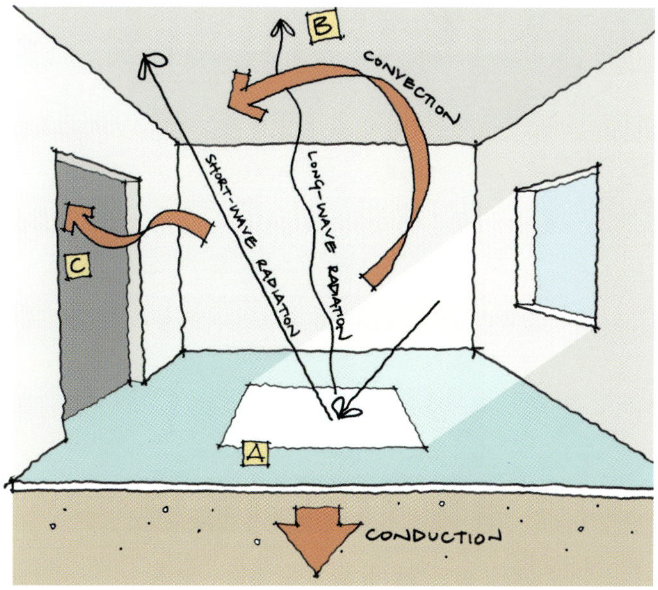

Figure 4.12: Solar gain delivered to thermal mass by (A) direct radiation and conduction, (B) convection and re-radiation, (C) convection only (thermal mass located in floor, ceiling and adjacent room respectively).

Thermal mass

Thermal mass is a term used to describe the property of the building fabric to respond to a given heat input with a temperature change. A building of low thermal mass will heat up quickly in response to solar gains, or auxiliary heating, and cool down quickly when they are absent. A building of high thermal mass will heat up and cool down slowly. The word *mass* is used because massive or dense materials, such as brick or concrete, also have large heat capacity.

The provision of thermal mass is often regarded as a passive strategy to improve thermal comfort. Heat gains to the interior of a well-insulated room will result in a rapid temperature rise unless there is thermal mass available to absorb some of the heat. This rise may take the room temperature above comfort level.

Location of thermal mass

A very important requirement is that the thermal mass is *coupled* to the heat gain. By this, we mean that it can receive heat by conduction, convection or radiation – as shown in Figure 4.12. In practical terms, if thermal mass is being used to reduce temperature swings in a room due to solar and/or occupant gains, then the mass must be inside the room and have good coupling – e.g. a large surface area – Figure 4.13.

Thermal mass is also used to increase the utilisation of solar gains, as shown in Figure 4.14. The principle here is that heat that may be delivered when there is no demand for heating would have to be rejected by ventilation or shading, unless it can be stored. If there is thermal mass available, the stored heat will be released passively when the room temperature drops below the temperature of the thermal mass, and will reduce the level of net auxiliary heating required.

It is not essential that all three mechanisms of coupling are present. For example, a suspended ceiling could be left partially open to allow air circulation between a massive floor slab and the room below, possibly assisted by a fan. Other advanced approaches include fan-driven airflow through ducts within the floor slab.

There follows a general rule, that in most cases thermal insulation is better placed outside the mass. This leaves the mass accessible to the interior (Figure 4.15). In cold conditions, where solar gains are useful, external insulation reduces the heat loss from the mass.

Thermal mass is not always good news. In a space that is intermittently occupied and intermittently heated, thermal inertia will require an earlier switch-on time, and will lead overall to higher energy consumption. This means that there may be some room types that are better off with fast thermal-response times, if it can be certain that they will not receive large thermal gains.

Lightweight construction

The increasing use in house construction of timber and timber-panel products, and of prefabricated structural insulated panels (SIPs), results in many buildings being short of thermal mass. This construction approach, which in many other respects is regarded as environmentally positive, can return satisfactory thermal performance provided a strict control is kept on casual and solar gains. Quite low-mass elements, such as plasterboard and timber, and room contents such as furnishings do have some thermal-storage function, due mainly to their large surface area. Even the air in a 4 m x 4 m x 3 m room has the thermal capacity of 60 kg of

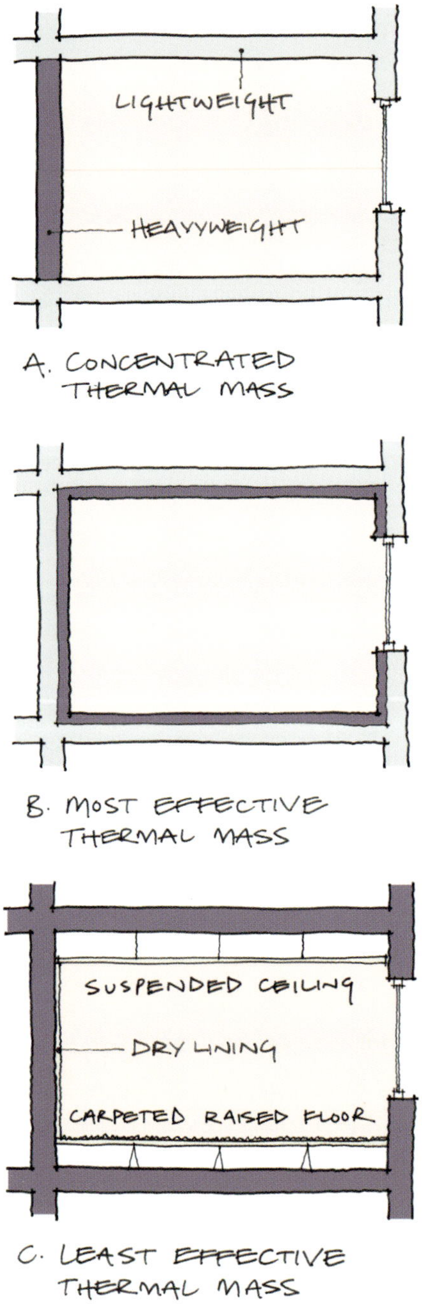

Figure 4.13: The location and distribution of thermal mass has a strong impact on its effectiveness.

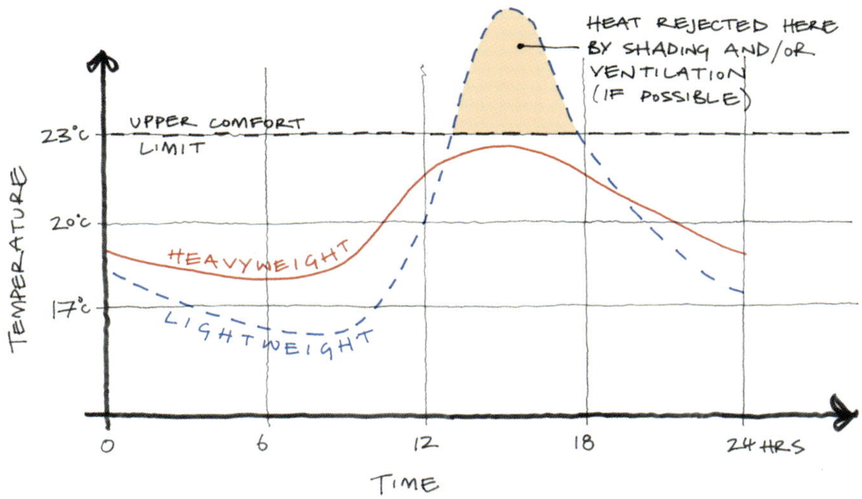

Figure 4.14: The impact of thermal mass on the utilisation of solar gain.

concrete, and can very quickly absorb a local heat input due to convection.

However, it is essential that solar gains in particular are controlled by effective shading, and large areas of solar-orientated glazing should be avoided. Furthermore, ventilation as a means of temperature control becomes more important in lightweight structures.

Moisture control and condensation

Moisture management and its impact on air quality will be dealt with in chapter 5. Here, we are concerned with interaction with the envelope and the prevention of surface or interstitial condensation.

It may come as some surprise to learn that there is a net flow of water from the interior of a house to the outside that, typically, amounts to nearly two tonnes per year. This is a product of respiration and of occupant activities such as cooking and washing. The majority of this leaves in the form of water vapour, by infiltration and ventilation. Some of it leaves by diffusion through the envelope (if permeable), although it is now realized that this mechanism accounts for only a small proportion of the total.

The physics of water vapour

The molecules of a liquid are in a state of motion and some have sufficient *kinetic energy* (in the form of random motion) to escape the surface of the liquid into the air above. This process is called *evaporation* and it leads to a net loss of energy from the parent liquid, which gives rise to a cooling effect.

Figure 4.16 illustrates two cases. Firstly, the molecules of the water are allowed to escape from the bottle. This would lead to the eventual evaporation of all of the water. In the second case, with the bottle closed, the concentration of water molecules in the air above the liquid water builds up until there are the same number of molecules re-entering the liquid surface, as leaving it. Thus, no further evaporation is taking place and the mixture of air and water is said to be *saturated*.

Bearing in mind that the kinetic energy of the molecules is related to temperature, it is not

HOUSE DESIGN

Figure 4.15: Thermal mass can be particularly effective during cold weather when it is located where solar and casual gains can be stored (in this case, in the middle of the house) until they become useful in the evening.

surprising to find that the amount of water molecules that the air above the liquid surface can hold – the 'saturated moisture content' – is also dependent on temperature. Figure 4.17 shows a graph of saturated vapour content as a function of temperature. We can see that air at 20°C can hold four times as much as air at 0°C, and it is this property of air and water that causes us the problem of *condensation* (see below).

Due to infiltration and ventilation, air inside a building is rarely saturated. Its state is described by the term *relative humidity*, which is defined as the mass of water in the air divided by the mass of water if the air were saturated at the same temperature. It is usually quoted as a percentage, e.g. 65% RH. Typical values to be found in a building range from 30% to 90%.

Diffusive and advective flow

Flow by diffusion is caused by the random motion of molecules: wherever there is a gradient in the concentration of a particular type of molecule – e.g. water – there will be a net flow down the concentration gradient. Advective flow is caused by the difference of pressure leading to the bulk flow of moist air; in the context of buildings, this is caused by thermal buoyancy or wind pressure.

A material such as plasterboard is porous at a microscopic level, and will allow water molecules (which are very small) to diffuse through it. However, bulk flow (advective) of moist air would be almost completely prevented due to the friction of the air flowing through the tiny pores. A material such as loose mineral fibres will allow both diffusive and advective flow, whilst an aluminium-foil or polythene membrane will virtually prevent even diffusive flow.

Condensation

We are now in a position to understand the problem of condensation. Consider the air of a living room at 22°C and 75% RH, which contains 12 g of water per kg of dry air. Assume that this air has free access (by diffusion and bulk flow) to the surface of a single-glazed window where, with an outdoor temperature of 3°C, the temperature of the inner surface is 11°C. Referring to Figure 4.16, we can see that air at 11°C can only hold about 8 g of water per kg of dry air; the surplus moisture

Figure 4.16: Evaporation of a liquid into (a) an open space and (b) an enclosed volume.

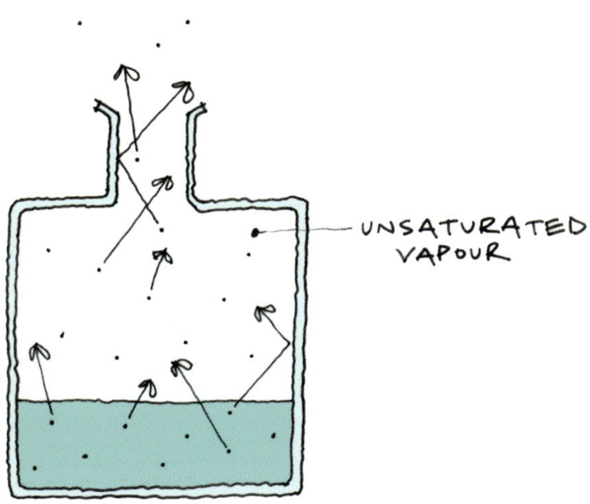

a

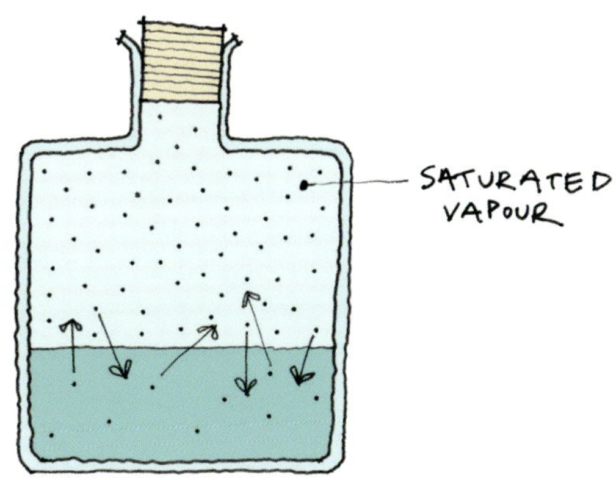

b

has to leave the air in the form of condensation on the surface of the glass.

This effect is called *surface condensation*, and can take place on poorly insulated opaque surfaces as well as glass. It can lead to physical degradation of the surface and mould growth, and is a common failing in older buildings that have been partially upgraded by reducing infiltration but leaving uninsulated parts to the envelope – e.g. solid masonry walls.

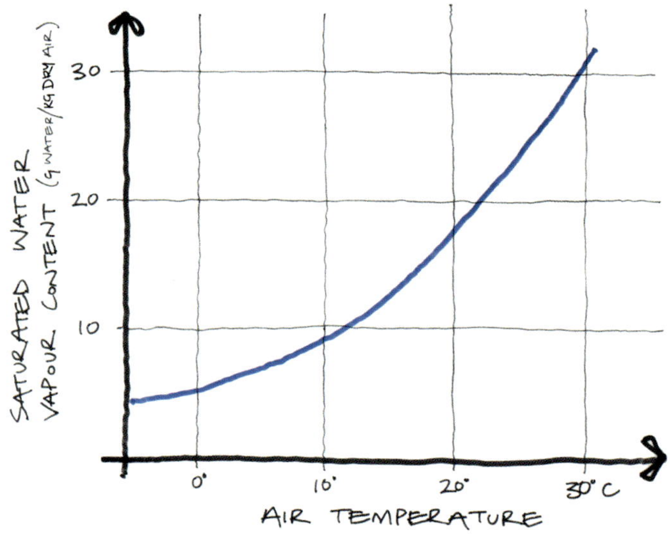

Figure 4.17: Saturated water vapour content for air as a function of temperature.

Now consider another case. The same room air is permitted to flow into the timber- framed wall shown in Figure 4.18(a). Because the fibrous insulation offers no resistance to airflow, and due to poor construction, it can gain access to the cold inner surface of the cladding. Because this is below the temperature at which the air becomes saturated (referred to as the *dewpoint*), condensation occurs. Condensation that takes place within the structure is called *interstitial condensation*, and can lead to serious structural damage.[9]

Figure 4.18(a) shows the general principle of interstitial condensation, in which the temperature gradient through the insulation drops below the dewpoint. The moisture content, and hence the dewpoint, remains almost constant through the inner layer of plasterboard and the insulation because, being porous, they offer no resistance to the diffusion of water molecules, unlike the outer leaf of the wall. Figure 4.18(b) shows the principle of applying a *vapour-check* membrane to prevent the migration of water by diffusion, and thereby lower the dewpoint in the cold part of the structure. This membrane will also help to prevent the advective flow of indoor air, driven by thermal and wind-induced pressure differences, to cold parts of the structure.

However, this is not to say that a vapour-check membrane is the only way to solve the problem. It is a matter of the *relative* permeability in relation to the temperature gradient. An inner lining of painted timber or plywood can provide sufficient vapour resistance relative to the freely ventilated outer cladding. If on the other hand the inner lining were penetrated in several places by poorly sealed services, the bulk flow of indoor air into the structure would lead to condensation. Or, if an impermeable membrane had, in error, been placed behind the outer cladding, this also would cause interstitial condensation.

Where it is necessary to reduce the bulk flow of air through a structure (in order to prevent the transport of moist air and/or to prevent heat loss by infiltration), a *breather membrane* should be used, as in Figure 4.18(b). This can provide high resistance to airflow and wind-driven rain, but allow water vapour to pass through by diffusion.

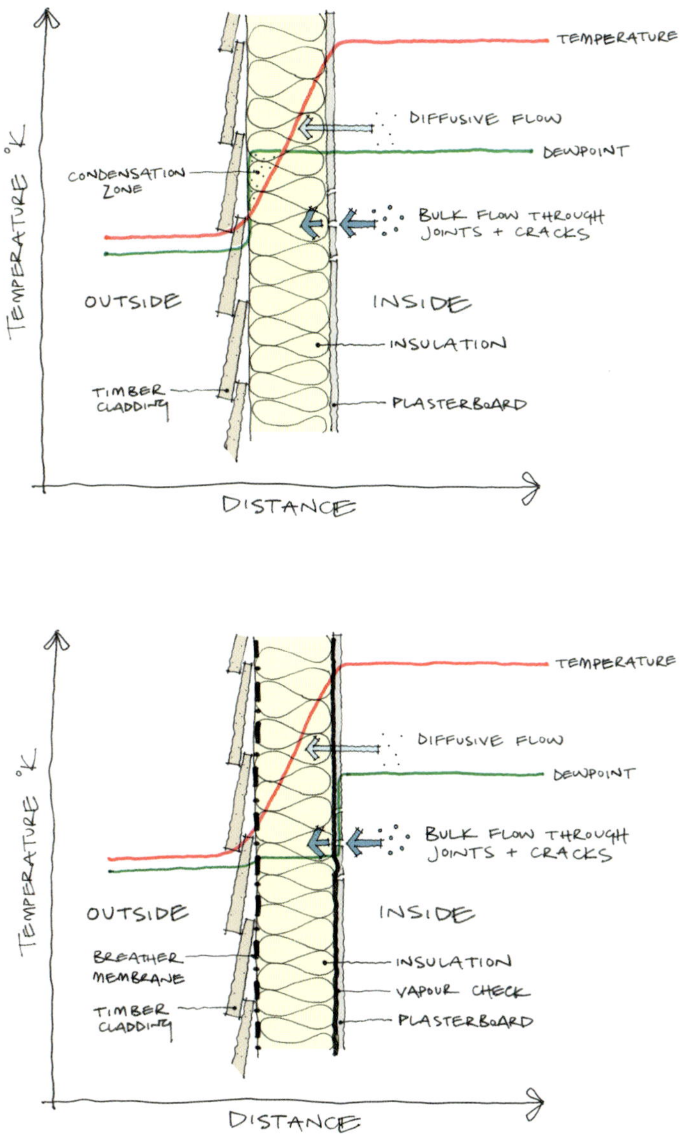

Figure 4.18: Dewpoint profile (green line) within a structure: (a) interstitial condensation with no vapour check, (b) use of vapour check to control dewpoint profile. The dewpoint is the temperature at which condensation will occur.

'Breathing' walls

This term has been rather optimistically applied to the concept that walls can be designed to allow air, water and other pollutants to diffuse through them without leading to interstitial condensation – or, at least, where some transient interstitial condensation does occur it is in part of the structure that will not suffer degradation from water.

Various claims have been made concerning the thermal performance of 'breathing' insulation. There is some evidence that the insulating effect of a porous material such as mineral wool shows a slight increase (about 10%) when airflow is from the cool side to the warm side. But there is a drop in insulation value when the reverse applies. Furthermore, tests carried out on a full sized 'breathing' building through a winter season showed that the majority of the airflow was still through cracks and via deliberate ventilation.[10]

The fact is that *all* buildings have to breathe – i.e. replace internal polluted air with unpolluted outdoor air. The question is how the air gets in and out. It seems that relying on air exchange solely by diffusion through a porous wall is quite impractical and does not carry demonstrable energy or health advantages. Indeed if we consider that the porous matrix could harbour pollutants or even pathogenic organisms, which would be difficult if not impossible to purge, it seems that ventilation via localized and controllable openings, is definitely preferable – *build tight, ventilate right*.

The comments above do not preclude advantages in certain 'breathing' products such as paints and renders, which relate to the mechanical properties of the element – e.g. the adhesion of a paint film. These products are increasingly being referred to as 'vapour-permeable' in order to distinguish them from the 'breathing envelope' concept.

Internal heat gains and energy balance

Heat gains to the interior of a room occur from three sources:

1. solar gains via windows
2. occupancy gains from lighting and other heat-generating equipment (also referred to as casual gains)
3. occupancy gains from human occupants (also referred to as metabolic heat gain).

Solar gains

Solar gains occur when any daylight or direct sunlight passes through the glazing of a room. Because diffuse daylight is typically 5 to 10 times weaker than direct sunlight, its heating effect is much smaller, so we tend to associate solar gain with just direct sunlight. Calculation of solar gain and heat balance were discussed in 'Heat loss by conduction' and 'Heat loss by ventilation' earlier in this chapter.

Source	*Gains (W)*
Occupants seated per adult	*90–120*
Television small/large	*50/150*
LED lighting per 100 lux per m²*	*2*
Tungsten lighting per 100 lux per m²	*15*
Laptop computer	*35*
Vacuum cleaner	*1,400*
Oven	*2,500*
Coffee maker	*1,200*
Hairdryer	*1,500*

** SI unit of illumination*

Table 4.3 Typical heat gains from room contents.

Occupancy gains

Table 4.3 shows various values for internal gains (in watts) from household equipment and activities. It must be borne in mind that equipment that is used intermittently will have average outputs far less than its rated peak output. Values are indicative only.

When room temperature is raised to within the comfort zone by solar and occupancy gains, there is no need for additional heating. If total gains are insufficient, the discrepancy has to be made up by auxiliary heating. Where the heating system has thermostatic controls, the auxiliary heat input will be adjusted automatically so that the total heat input just maintains the set temperature. Poor or absent thermostatic control can lead to over- or underheating and, hence, discomfort.

Overheating is often dealt with by increasing the ventilation rate, but where this is insufficient mechanical cooling can be applied. This requires extra plant and represents a further use of electrical energy. There is growing interest in reducing the use of mechanical cooling by passive means – i.e. by the inherent properties of the building. This involves the use of shading, thermal mass and natural ventilation.

Key points

- Insulation
  - reduces heat load in the heating season and can reduce cooling load in the cooling season in air-conditioned buildings
  - can lower peak indoor summer temperatures in a free-running building[11] if located in an element, such as a roof, that has a high surface temperature due to insolation.

- Thermal mass
  - can reduce temperature swings from uncontrolled gains and limit peak temperatures, particularly in combination with night ventilation
  - can increase the utilisation of solar gains in the heating season
  - can increase the energy consumption in intermittently heated buildings.
- There must be good coupling between thermal mass and thermal gain.
- In buildings with intrinsically low thermal mass, strict control of solar gains by shading, and ventilation, is essential.
- Warm air can hold much more moisture than cold air. Condensation occurs when warm moist air is in contact with a cold surface or porous material.
- Interstitial condensation
  - occurs within the structure and can be prevented by the use of a vapour-check membrane on the warm side of the insulation
  - can also be prevented by ensuring that the relative permeability of the structural layers prevents the dewpoint from dropping below the temperature of the structure.
- The bulk flow of air can also conduct moist air to cold parts of the construction, and may be more in need of control than diffusive flow.
- Breather membranes on the cold side of the insulation can be used to prevent bulk airflow and wind-driven rain, without reducing vapour permeability.
- House ventilation by diffusion through permeable walls is not practical and carries no significant environmental benefits.
- Local thermostatic controls reduce the risk of overheating and reduce energy consumption.

Noise and its control by the building envelope and structure

Noise is a major factor in the wellbeing of occupants. Sources of noise include aircraft, industrial activity, road traffic and noise generated by neighbours. To some extent noise nuisance can be reduced by site planning, as has been described in 'Urban noise and its mitigation' in chapter 3. Here, we consider the role of the building envelope, including party walls, in reducing unwanted sound.

Principles of sound reduction

We have already dealt with the reduction of sound with distance in 'Urban noise and its mitigation' in chapter 3. Here, we are concerned with the passage of sound through a structure.

A sound wave is a series of high- and low-pressure zones that move through the air. When the waves meet a partition (whether plasterboard, brick or even fully glazed), the high-pressure zone 'pushes' the partition and the low-pressure zone 'pulls' it, causing it to vibrate. The vibrating partition then affects the air on the other side, alternatively compressing it and de-compressing it to form another progressive sound wave, as shown in Figure 4.19. It follows that the stiffer the partition, or the more inertia (due to its massiveness) that it has, the less it will vibrate in response, and the weaker the transmission of sound will be.

This is the basis of the mass-loss law, which relates the Sound Reduction Index (SRI)[12] to the mass per m^2 of a partition, as shown in Table 4.4.

Note that the SRI of a given partition is greater for high frequencies than lower, particularly for high-mass brick partitions. This is fortunate in respect of privacy between rooms because most of the information of speech is carried in the higher

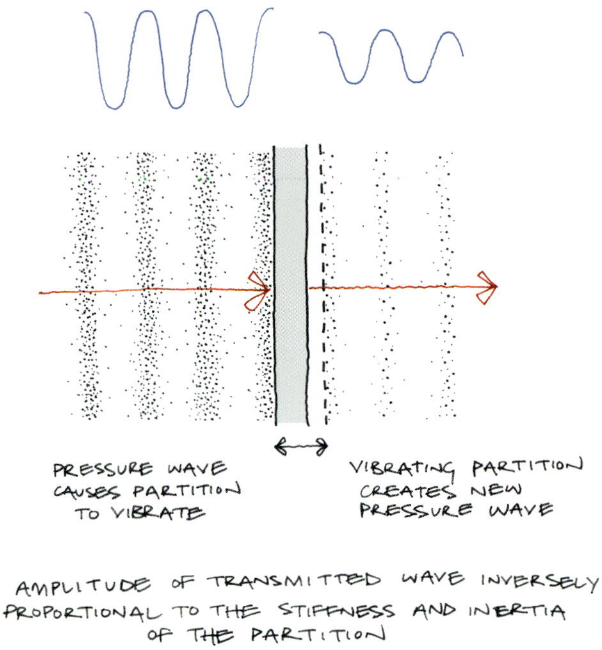

Figure 4.19: Transmission of sound waves through a partition.

frequencies. Thus, hearing a low unintelligible murmuring from our neighbour is generally acceptable.

Sound transmission by multiple layers

Because of the non-linear nature of our perception of sound, the performance of two independent partitions is better than one of twice the mass – i.e. their SRIs are additive. However, this is only true if Partition A is completely structurally independent of Partition B. If, for example, the two are of masonry joined with cavity ties or two layers of plasterboard fastened to the same timber stud, as in Figure 4.20(A), the layers will act as a single structure. Figure 4.20(B) shows a partition designed to separate the two layers, and provide porous material within the cavity to absorb sound reflecting back and forth within the cavity. This partition will have an SRI approaching the sum of the SRIs of the individual plasterboard layers. This demonstrates an important principle in noise control – the role of the structure itself in transmitting sound.

Description	*Mass/m²*	*Sound Reduction Index (dB)*		
		250 Hz	*500 Hz*	*1,000 Hz*
Glazing, single 3 mm (sealed)	*7.3*	*16*	*23*	*26*
Glazing, double 4 mm	*21.6*	*19*	*27*	*31*
Stud partition: 12 mm plaster-board x 2, on timber studs	*26*	*34*	*36*	*42*
Brick, 100 mm	*200*	*34*	*39*	*46*
Cavity wall: brick 100 mm x 2, cavity	*410*	*38*	*49*	*63*

Table 4.4: Dependence of Sound Reduction Index on mass of partition and frequency of sound (for illustration only).

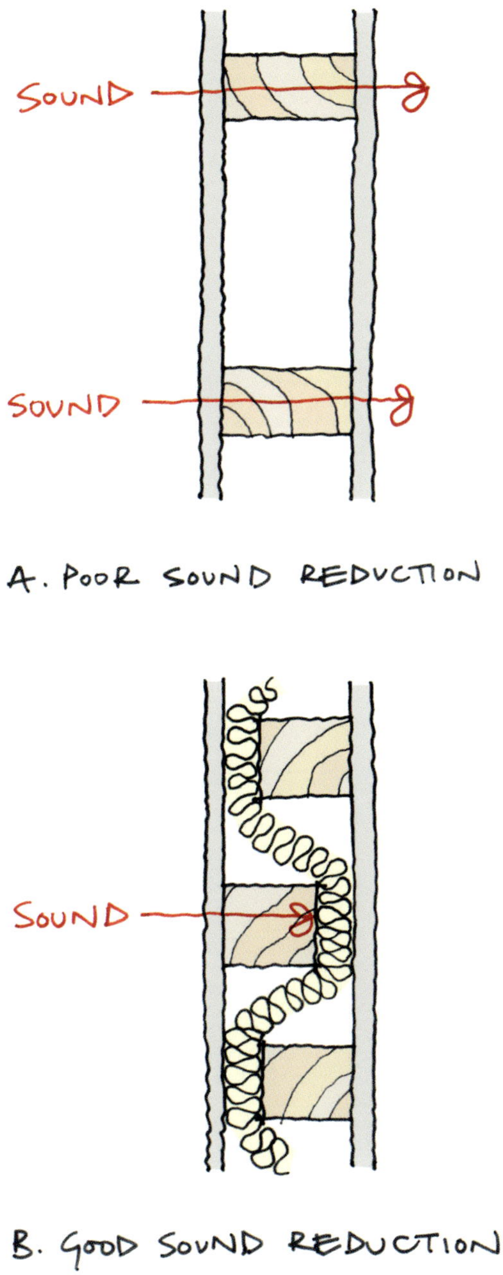

Figure 4.20: Structural design of stud partition with: (a) poor SRI and (b) good SRI.

Effect of gaps on the sound reduction of a partition

One issue that is a common cause of failure in noise control is the effect of perforations, gaps, cracks, etc. in a partition. Bearing in mind that sound is a pressure wave, it is not surprising that during the positive part of the pressure cycle, air will flow through a gap, and vice versa. This air becomes a new sound source as it emerges on the 'quiet' side of the partition, as shown in Figure 4.21 – and it has been uninfluenced by the inertia and stiffness of the partition.

It follows from the physics that a small gap in a partition causes a disproportionately large reduction in SRI, as indicated by Case B in the graph in Figure 4.22. This graph also shows the general case in which sound passes through two elements in parallel – e.g. a window and an opaque wall (Case A – see *Use of SRI graph* below).

Gaps in partitions caused by poorly fitted service pipes or ducts, or poor-quality construction, are frequently to blame for noise problems. This is especially so in existing buildings, where critical sound paths are hidden by other internal structures and fittings. Indeed, it is recommended that a sound-reduction test be carried out in order to detect unexpected weaknesses before any structural alterations are completed.

Use of SRI graph (Figure 4.22) for composite structures:

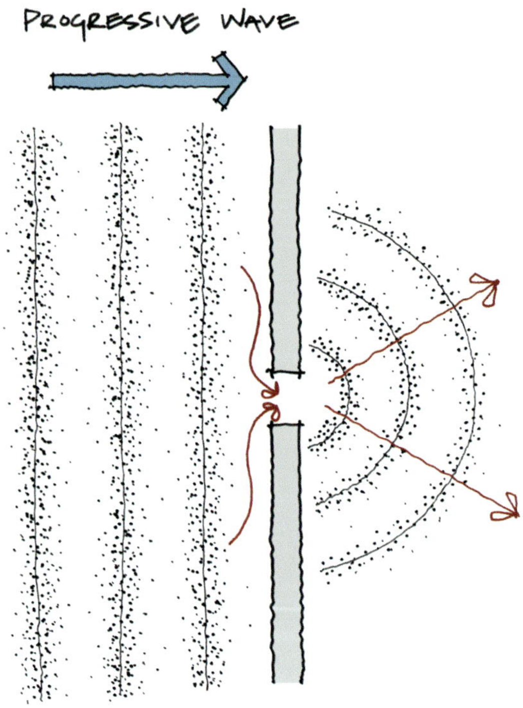

Figure 4.21: Air flowing back and forth through a gap becomes a new sound source.

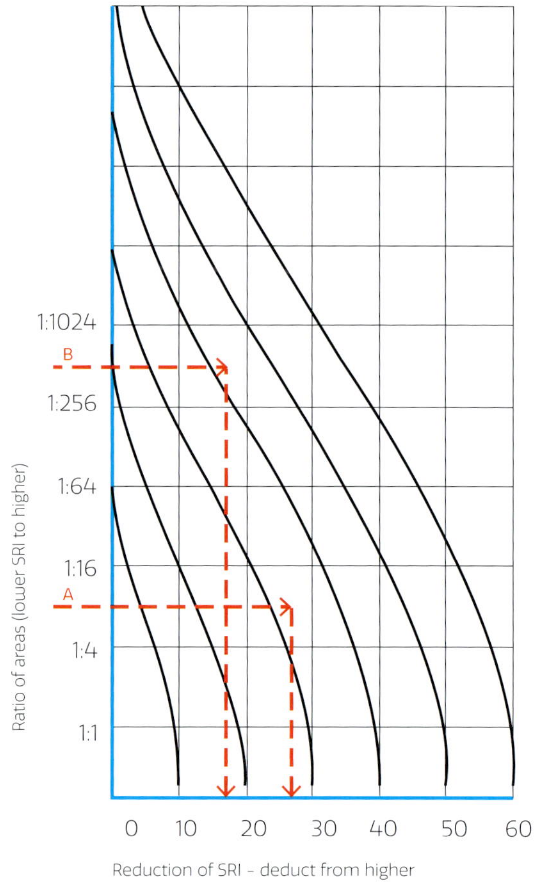

Figure 4.22: Graph showing SRI for combined areas with different individual SRI.

Case A

A brick cavity wall (SRI 65 dBA) measures 4 m x 2.3 m and has a double-glazed window (SRI 31 dBA) of 1 m^2. The ratio of areas is 1:8.2 and the difference is 34 dBA. Enter the Y-axis at 1:8 and move horizontally to the (interpolated) 34 dBA difference curve. From the intersection, drop vertically down to the X-axis at 28 dBA. This is deducted from the higher SRI, 65 dBA, giving an SRI for the wall/window combination of 37 dBA.

Case B

A stud partition has an SRI of 42 dBA. Holes for services and other work result in a total gap area to partition area ratio of 1:500. The gap has zero SRI, so the difference is 42 dBA. Using the graph as above gives a resultant value for the partition of 42 – 17 = 25 dBA.

Flanking sound

Another issue that affects the apparent performance of a partition is flanking sound. This is where the sound-reduction performance of a partition is effectively compromised by the sound taking an adjacent route that has a lower SRI. This has already been illustrated in Figure 3.11 in the previous chapter. It is a common problem in buildings in which alterations have been carried, such as rooms being divided, and in older terraced and semi-detached houses where two units may share a common roof structure.

Another route for flanking sound is via ventilation ducts. This will rarely occur between different households but could take place between rooms, when extract ducts from a kitchen or bathroom pass through the ceiling void of another room. The growing popularity of mechanical ventilation with heat recovery (MVHR) will inevitably increase the chance of this problem occurring. It may be solved by re-routing the duct; a more technical solution, employing sound attenuators in the ducts, would require specialist input.

Impact and structure-borne sound

Airborne sound interacts with a structure (partition, floor or envelope) in the way described above in 'Principles of sound reduction'. In that case, we were primarily concerned with sound sources that deliver their sound energy to the air – including speech, musical instruments, and domestic appliances such as a vacuum cleaners and washing machines.

In the case of impact sound, the energy is delivered direct to the structure: footfall on the floor above, a door slamming, or vibration from a fan or pump. Once the sound wave is set up in a material, it is attenuated much less rapidly in strong materials such as steel, concrete and even timber than in air. It may travel long distances; we are probably familiar with the sound of a

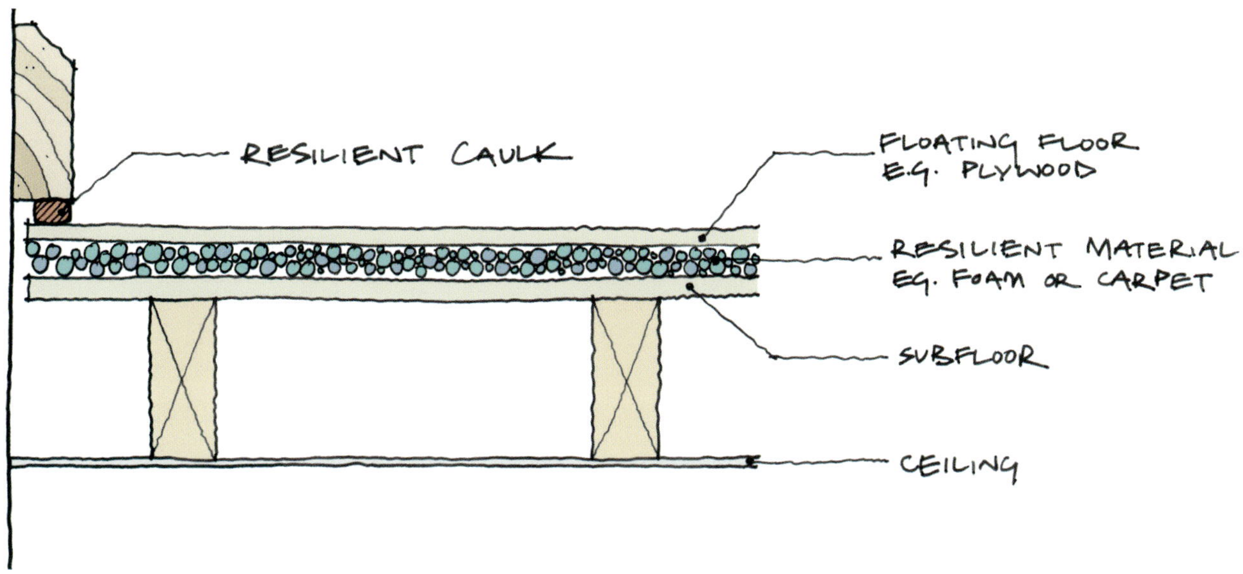

Figure 4.23: Simple floating floor to reduce structure-borne sound between rooms.

masonry impact drill 'somewhere in the building'. The problems of impact sound must not be underestimated, particularly in housing. In one investigation carried out by the author, the main cause of annoyance from a neighbouring attached property was the sound of a toilet-roll holder revolving and then banging against the party wall!

The solution to the problem is to prevent the sound energy from getting into the structure in the first place. In the case of impact sound on floors, the use of a floating floor – a rigid deck supported on a resilient material and not rigidly attached to the structural floor – will provide a solution, as in Figure 4.23.[13] Note that if the deck is rigidly fastened through the resilient material the impact vibration due to footfall will be transmitted to the structural floor, and its effect negated. Or more simply, a resilient covering such as a carpet will reduce the impact vibration.

In the case of machinery noise, the best solution is to use resilient mountings. These can be 'tuned' to have the best frequency response to the vibrations, and specialist advice should be sought in their design.

Sound absorption and reverberation

Sound absorption

There is often some confusion between sound *absorption* and sound *insulation*. If it is light and fluffy and porous, then it is not sound insulation; however, it may be good *thermal* insulation – and this is where the confusion lies. The author has seen advertisements for 'sound insulating wallpaper' – a sheet material with a foam backing 3 mm thick. This would have absolutely zero effect on the SRI of, for example, a masonry wall.

Sound insulation is, in fact, not the property of a material at all – it is the property of a structure that separates a sound source from a receiver, and indicates the reduction of sound as it passes through that structure. Acoustically well-insulated structures have to be heavy (inertia) and rigid, so that they do not respond to the pressure fluctuations. And they must not be leaky, allowing a new pressure wave to be generated by the 'leak'.

Sound absorption occurs where sound is converted into heat by the process of friction, as the air flows in and out of a porous material. Sound absorption is most significant when it occurs in the same space as the sound source and the listener. This is because even a good absorber (e.g. 25 mm of porous fibreboard) will only absorb about half of the sound energy, at best. Due to the way in which we perceive sound, this would only reduce the sound level by 3dB, which would be almost undetectable to the ear.

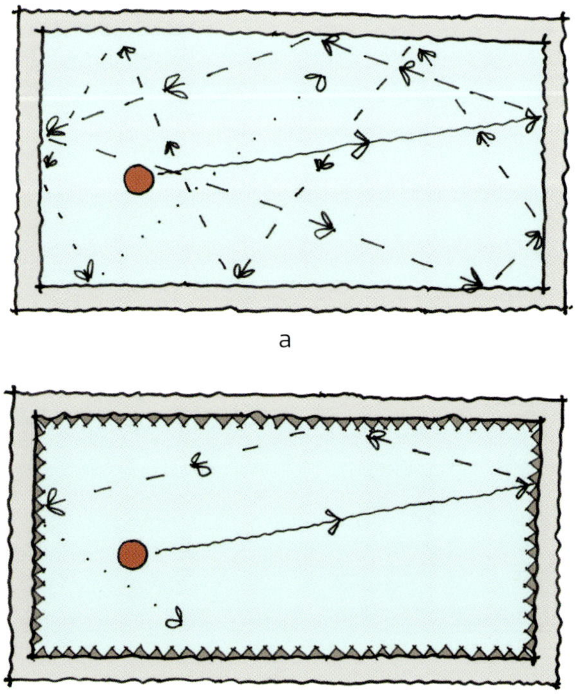

a

b

Figure 4.24: Multiple reflections in a room with (a) no absorption and (b) with absorption.

Reverberation

Figure 4.24(a) shows the effect of a sound generated in a room with very little acoustic absorption. The sound reflects many times from the hard surfaces of the wall, adding to the loudness perceived by the listener. Furthermore, because of the finite speed of sound through air (343 m/s), the later reflections are out of phase with the early one – and this, particularly if the sound is speech, reduces its intelligibility. This effect is known as *reverberation*. In the case of music, reverberation can enhance the perceived quality of the sound to a certain extent.

If the sound source (the red circle in Figure 4.24) is suddenly cut off, the reflected sound will continue to bounce around the room until so many reflections have taken place that it becomes inaudible. The time taken to do this is known as the *reverberation time*.

If we now cover the walls with an acoustic absorber, as in Figure 4.24(b), the multiple reflections will die down much quicker because at each reflection a much greater proportion of the sound energy will be absorbed. The number of overlapping reflections will be reduced, as will the reverberation time. Speech will sound clearer, but the room now may be too 'dry' for music.

Design for optimum reverberation in relation to room use, and the distribution of reflective and absorptive surfaces to ensure sound distribution, is a specialised subject of room acoustics. It mainly has application in the design of classrooms, lecture theatres and auditoria, and is rarely an issue in housing design.

However, sound absorption does have a role in sound insulation. Figure 4.20 showed a partition that includes a cavity in its stud zone. Because each of the layers of plasterboard is independently supported, reflections would take place between them and the sound level would build up unless an acoustic absorber was included in the cavity. Even in a conventional studwork wall (Case A), there will be some improvement by filling the cavity with a mineral wool – but it should be of the highest density available. In a similar way, reducing the reverberation in a room by absorption will lower the overall sound level within that space, and therefore result in less sound transmitted to an adjacent room. Note that in both these cases, the absorptive material is in a cavity in which multiple reflections can take place.

Design responses

In designing the building envelope, or even the envelope of a room, the principle of a chain being as strong as its weakest link is a good one to follow. Thus, if a well-constructed bedroom has a badly fitted door onto a noisy corridor, efforts to improve the SRI of the walls will be largely wasted. Similarly, if a façade with openable windows faces a noisy road, effort should be applied to improving the window performance rather than that of the masonry wall. Finally, don't waste your money on 'sound-insulating' wallpaper!

In most cases, using the principles described in the preceding paragraphs, noise nuisance can be mitigated by the design of the building envelope.

Key points

- Neighbourhood noise nuisance is a major issue in dense urban housing.
- The design of the building envelope can help to mitigate noise nuisance.
- Heavy and/or rigid separating structures are more effective at reducing sound than lightweight flexible ones.
- The sound-insulation value of a composite element in which sound passes through two different structures in parallel tends to be dominated by the lower value.
- Airborne sound can pass through cracks and gaps, with an effect disproportionate to their actual area.
- The sound reduction of a partition can be apparently reduced by sound following a flanking route other than through the partition itself.
- The sound reduction of a two-layer wall or partition approaches the sum of the two individual layers' reductions if they are mechanically independent and there is sound-absorbing material between them.
- Sound-absorption materials reduce reverberation in rooms.
- Sound-absorption materials contribute to sound reduction only when in a room or cavity where multiple reflections can take place.

Chapter 5

The building interior and services

Introduction

This chapter examines in detail the following aspects of the building's interior:

1 Thermal comfort
2 Ventilation
3 Natural light
4 Artificial light
5 Noise

Environmental Comfort

Of the various environmental parameters that affect our comfort, temperature is probably the most critical, and one that we expend most effort to control. This is because the range of temperatures that the outside environment covers, on a daily and seasonal basis, frequently includes conditions that are not only very uncomfortable but possibly life-threatening.

During the latter part of the 20th century, the conventional wisdom was that thermal neutrality should be the target. It seemed very reasonable that people are comfortable when they are neither hot nor cold. Experiments were carried out in climate chambers by Fanger[1] to develop a theory based on instantaneous heat balance – whereby the heat gain exactly equalled the heat loss to the environment. This led to a neutral temperature, which was used to establish international temperature standards. It was upon the perceived need to achieve these parameters, in both the work place and at home, that the air-conditioning industry was built.

Adaptive-comfort theory

However, there was conflicting evidence. It was pointed out by Michael Humphreys[2] as early as 1975 that when comfort surveys are carried out in the home and workplace instead of comfort chambers, people report comfort over a much wider range than that predicted from Fanger's heat-balance theory. Humphreys also noticed that the degree to which occupants can take *adaptive*

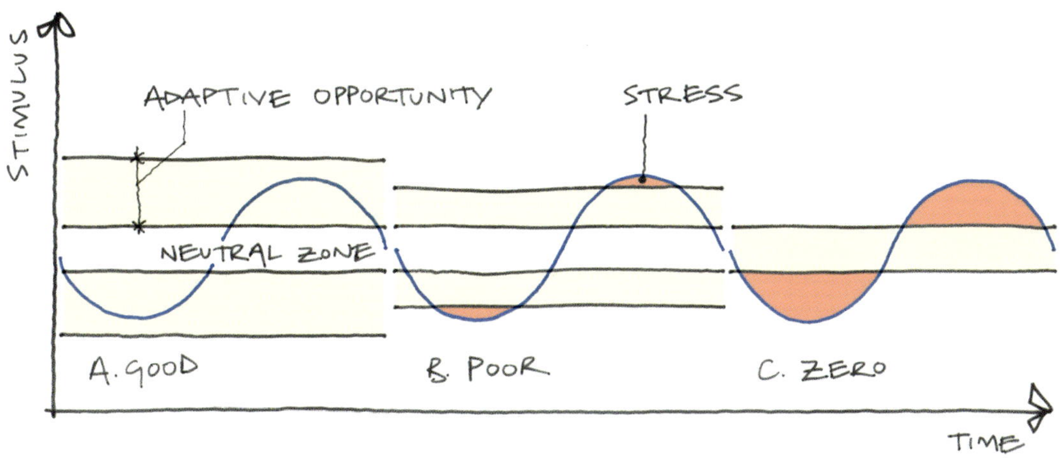

Figure 5.1: Impact of adaptive opportunity on depth of comfort zone.

action (such as opening a window or adjusting clothing) greatly influenced their satisfaction levels. This is illustrated in Figure 5.2, which shows a hypothetical response of an occupant to swings in an environmental parameter (such as temperature) away from neutral. It goes some way to explain why there appeared to be such a discrepancy between Fanger's findings in the climate chamber, where all *adaptive opportunity* is removed, and in the real world where in most cases some adaptive opportunity exists. Later work (Matias, et al.)[3] showed the *perceived* presence of adaptive opportunity to have a beneficial effect *even when the opportunity was not actually taken*. This suggests that there is a strong psychological component to environmental satisfaction.

What do we mean by adaptive action? Any action that reduces discomfort is referred to as adaptive. This can include adjustments to the environment – e.g. turning up the heating or opening a window – or adjustment to the person – e.g. a change in clothing, posture or level of activity. The term's biological definition refers to an action that ultimately increases an organism's chance of survival. This seems like a large step when compared with opening a window, but in fact the two definitions are linked.

In our everyday lives, we do not associate feeling a little too warm as being anything to do with survival. However, it is in fact an early-warning signal that if no action is taken and our thermal imbalance continues, our core body temperature will rise to danger levels and life will be threatened. In most cases, we do not let the situation get anywhere near fatal; however, there are people such as the sick, aged or the very young who are either incapable of adaptive action or, due to sensory malfunction, do not 'read' the discomfort signal – and in those cases, the results *can* be fatal.

Figure 5.2: There is something comforting and compelling about a real fireplace, despite the inconvenience and risks, even in contemporary homes that have efficient central heating.

In the workplace, individuals often feel that they do not have the freedom to make alterations to environmental controls that may be automatic or centrally controlled. Furthermore, they may be reluctant to make personal adaptations to clothing, posture and activity levels due to the nature of the job and its behavioural codes. The situation in the home is much more favourable for adaptive opportunity. In a longitudinal study,[4] Nigel Oseland[5] observed that people are more tolerant of non-neutral conditions in their home than in their workplace. The explanation is that there are more adaptive opportunities in the home and, importantly, that the occupants felt more able to use them.

However, there is still a deep-seated belief by some that the building should provide neutrality, and this can be achieved by ever-more sophisticated controls. Digital technology has made this irresistible to some designers, and the label 'smart building' has become a marketing tool. A really smart building would recognise the behavioural instinct of adaptation, and work with this, but all too often smart buildings take away control from the occupant. Furthermore, even when some level of control is permitted it is often via an interface that is neither ergonomic nor intuitive. Compare, for example, the action of opening a window because it is a little too warm with the action of addressing multi-function buttons or a touch-screen.

There are parallels in other environmental variables too. In lighting, many designers gave up on daylight as a working illumination, claiming that it was too variable and unpredictable. High levels of uniform artificial lighting became the norm, and daylight is now more to do with building aesthetics than function. This is largely due to a similar misunderstanding – the notion of uniformity and neutrality. Studies by Katerina Parpairi[6] have shown that occupants prefer daylight despite poor illuminance geometry and intensity, provided they have some adaptive opportunity to mitigate its negative effects. It was also shown that far-from-'ideal' lighting conditions were acceptable if they were seen to be due to 'natural causes'. This phenomenon can be observed when the occupant of a room lit by the failing evening daylight will carry on a visual task such as reading at illuminances as low as 50 lux, without switching the lights on. This is five times less than the recommended value that would be used to specify artificial-lighting levels.

Adaptive opportunities

What are adaptive opportunities, and can the designer provide enough – or even too many? In non-residential buildings where most studies have been carried out – typically, offices – the following positive adaptive attributes have been identified:

- a relaxed dress code
- workstation/furniture flexibility and occupant mobility
- access to hot/cold drinks
- openable windows and adjustable shading/blinds
- local heating/cooling controls and a desk fan or locally controlled ceiling fan
- shallow plan – minimising distance from window
- cellular rooms – reduces mutual disturbance
- daylighting with task-lighting backup
- good views – external and internal transition spaces e.g. balconies, atria – and access to outdoor areas.

Most of these could be expected in residential buildings. The question then becomes: can anything be done to increase their effectiveness or increase the probability of their use?

Take the fourth item, for example: if the only openable windows give onto a busy, noisy and polluted street, it is far less likely that they will be used than if they opened onto a quiet private or semi-private space. A good design response might be sealed windows onto the street to provide a view and useful light, and openable windows facing a protected area – even if the view was limited. This may make certain demands on room planning, and indeed the whole plan of the house or apartment.

An even better solution would be to provide openable windows with high Sound Reduction Indices (see 'Noise exposure' in chapter 4) on the street façade. This is because there almost certainly will be times when the street is quiet, and even when noisy, if the occupant feels that they have freedom of choice they will be more tolerant than if that choice has been predetermined.

What does all this mean for sustainable healthy design?

It is good news: it means that many design responses that are consistent with good energy performance (such as shading, natural ventilation and provision of daylighting) are also conducive to improved comfort and wellbeing, provided three basic principles are observed:

1. The global characteristics of the building design limit the range of environmental conditions – e.g. avoiding extreme temperature swings – by passive means.
2. Sufficient adaptive opportunities to cope with the environmental range must be provided.
3. Control systems should anticipate and encourage occupant intervention.

There is also another positive aspect when a number of environmental issues are addressed, which Adrian Leaman and Bill Bordass[7] refer to as the 'forgiveness factor'.[8] Based on data collected in many non-residential building-occupant surveys, they have found that although individual environmental parameters in certain types of building – e.g. temperature, air quality, lighting – may score only moderately well, when it comes to reporting the *overall* satisfaction with the building occupants record a higher score than the average individual scores. These buildings are nearly always those where control is visible and intuitive, and are predominantly passive rather than highly mechanically serviced buildings. It has not been proven that this response is found in housing, but studies of student hostels and hotels have suggested this to be likely.

This is consistent with the earlier assertion that occupants are more tolerant of non-neutral conditions when they are perceived to arise from natural causes – and we can add to that, when they understand the means by which they, the occupant, could mitigate them.

Extreme temperatures and health

A wealth of evidence links morbidity and mortality with exposure to both low temperatures experienced in cold winter spells[9,10] and high temperatures in summer heatwaves.[11] Excess winter deaths – i.e. those above the mean death rate for the year – have been recognised for decades and considered in relation to housing design and fuel costs. More recently, linked to the phenomenon of global warming and urban heat-island effects, there has also been growing interest in the impact of heatwaves.

Inevitably, the effects are most severe on the most vulnerable people – namely, the old, sick and very young. However, increased morbidity is observed in all ages – especially in very low-income families. Whilst many deaths are a direct result of hypo- or hyperthermia, extreme temperatures can have other morbid effects such as respiratory and cardiovascular diseases, and even mental-health problems.

The majority of casualties have occurred in old buildings, and bearing in mind the slow replacement rate of our existing building stock it is imperative that retrofit measures are given high priority. Good principles for new build should be applied to refurbishment projects as far as reasonably possible. However, there will be constraints – for example, it might not be possible to insulate an apartment block externally, thereby maintaining the benefits of the thermal mass, or a Modernist block may have its excessive glazing area protected by historical listed status. Where compromises have to be made, it is essential that the principles of good passive design are understood as well as the socio-economic status of the occupants and how it is likely to affect their behaviour.

The sociological and epidemiological significance of this topic has been introduced in chapter 1. Here, we concentrate on the technical and design issues of the building's fabric and services.

Impact of insulation, thermal mass, shading and ventilation on overheating

Improved insulation standards across Europe mean that heating costs in new or recent buildings are very low. This has all but removed the likelihood of severe underheating due to fuel cost. However, a well-insulated building can overheat, or be even more prone to overheating than a poorly insulated one. Much of Europe has a climate in which periods of high summer temperatures and low winter temperatures occur – thus the issues of controlling these extremes of temperature are best considered together.

- A fundamental rule of building science is that internally accessible thermal mass limits extremes of temperature (Figure 4.12).
- A second rule is that controlling solar gain by shading is an essential part of temperature control.
- Finally, ventilation plays a vital part in rejecting daytime heat gains as well as cooling thermal mass at night.

The correct manipulation of these elements can greatly reduce extreme temperatures.

However, the degree to which the occupant is required to make intelligent interventions with the building controls is of concern. Vulnerable people may find this difficult physically or due to reduced awareness,[12] or simply not understand the need. This strongly suggests that the inherently passive performance of a building is of great importance (Figure 5.3).

To attempt this strategy with fixed shading devices and fixed ventilation openings will seriously compromise overall performance. Rather, we should consider a design that requires minimal intervention – say a winter, spring, summer and autumn setting – and ensure that there is the human resource to make this adjustment. The limitation of this seasonal approach is that due to the random nature of weather – particularly in coastal western Europe, dominated by Atlantic weather systems – unseasonal episodes of significantly warm or cold weather often occur.

Alternatively, intelligent automatic control could be used. If so, care should be taken

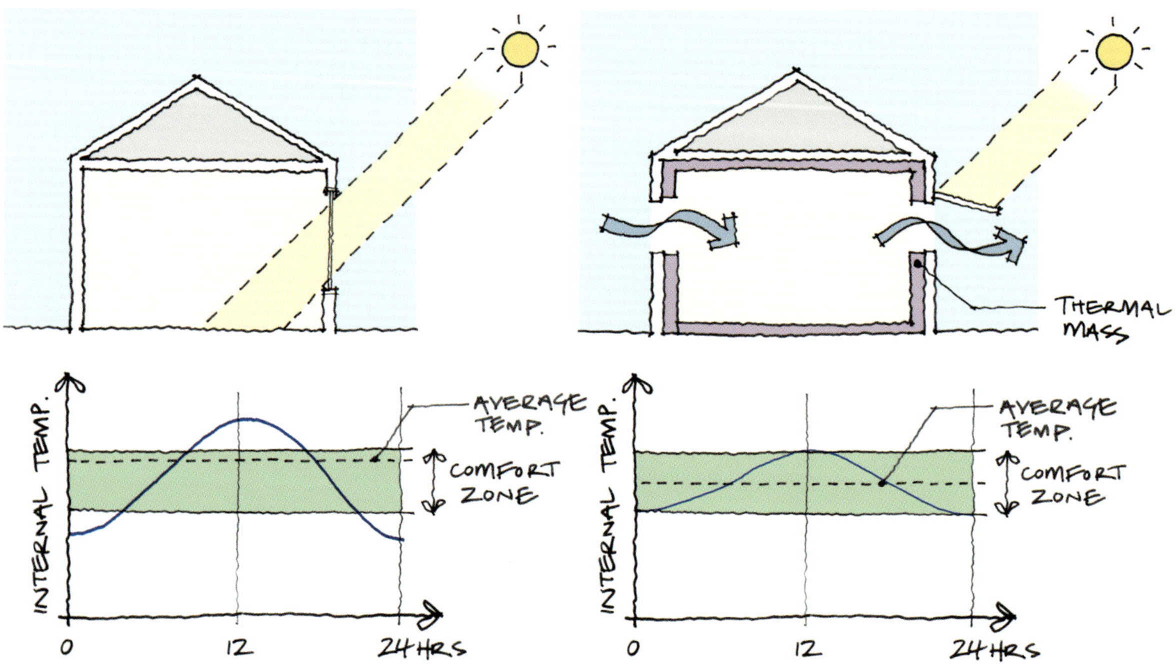

Figure 5.3: The impact of thermal mass, shading and ventilation on the passive response of a building.

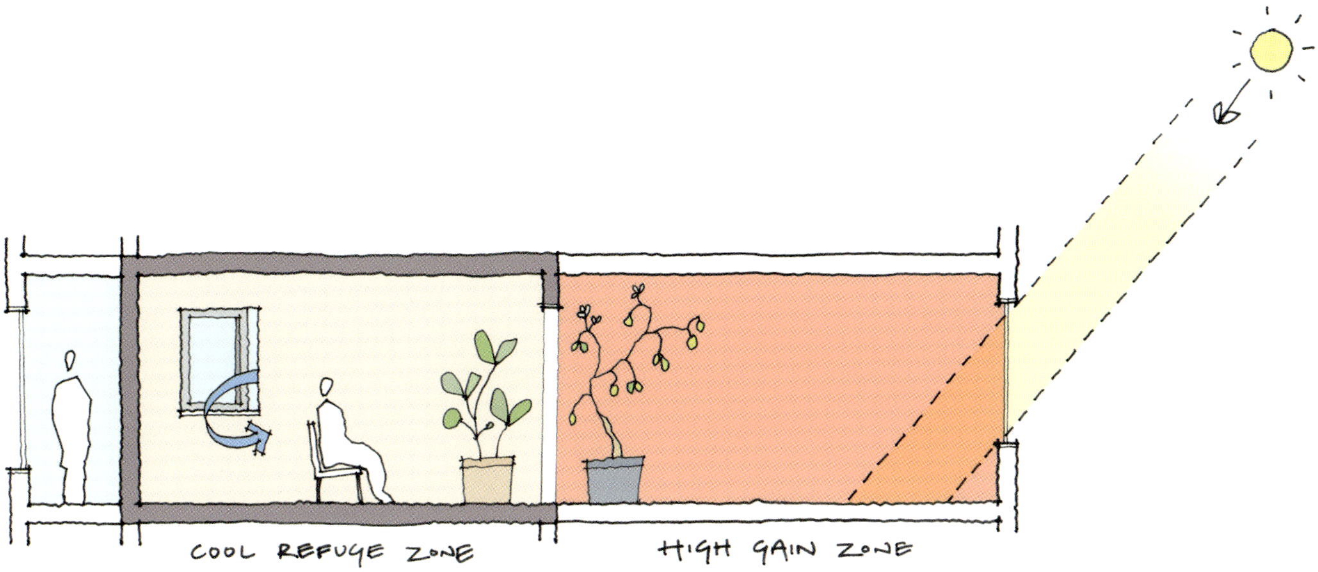

Figure 5.4: A cool refuge space away from solar gain, with ventilation and accessible thermal mass.

that the automated actions do not make sudden unpredictable changes to the sensible environment. For example, if motorised blinds are employed it might be better to have them deploy once at a fixed time during the day than rapidly deploying in response to instantaneous parameters such as sky conditions and temperature. As with any automated system, it is important that there is a failsafe and monitoring function, and that these can be manually overridden.

Spatial thermal design

The notion of environmental variance and providing adaptive opportunity within a space has been introduced in 'Adaptive opportunities' earlier in this chapter. This principle can be applied to extreme thermal conditions and vulnerable occupants. For example, if there is a risk of underheating due to health or economic status, then there should be a refuge space, perhaps a room or area in a room away from the main glazed area, which can be heated to a higher temperature whilst economising with lower temperatures elsewhere. It is essential that the heat emitters and control system allow this to happen and also critical that the occupant understands the reason for this extreme mode.

The provision of a cool refuge space may also be necessary. This is not as straightforward as the warm refuge, since few residential buildings have mechanical cooling. In small housing units without shared common spaces, this will be a challenge. Clearly the space must have minimal solar gains and accessible thermal mass, but it must also have ventilation access, preferably not from a sunlit outdoor or indoor space, as in Figure 5.4.

In older multi-occupancy buildings, serious and often fatal cases of hypo- and hyperthermia occur on top floors under poorly insulated roofs. This suggests that roof insulation should be given the highest priority in retrofit work. Rooms built in attic roof spaces often lack thermal mass for structural reasons, making the minimising of solar gain by conduction and through roof glazing even more essential.

Multi-occupancy buildings also offer possibilities. A shared common room or recreational area can be closely thermally controlled but have thermal variance within itself – i.e. in summer, a cooling-off area with lower temperature and higher air movement; or in winter, a cosy warm area with visible and/or touchable heat emitters. Once the attraction became known by residents, it would serve an important social function: isolation was found to be a contributing factor in many extreme-temperature deaths.[13] This would be especially appropriate in housing for vulnerable people.

The free-running building

As buildings have become better insulated and more airtight, total heat loss has dropped dramatically. This means that internal gains from occupants, appliances and solar gains can meet the heating load for much of the year. It is not uncommon to assess this by simulating the 'free-running' temperature without heating or cooling, to see what fraction of the year the temperature in the building lies within the comfort zone (Figure 5.5). It is an indication of the *passive performance* of the building. Note that the apparent performance improves when an adaptive comfort zone is applied – a truer indication of thermal satisfaction.

The simulation demonstrates that heat gains that are useful in winter may become unwanted in summer, causing overheating. To prevent this, assumptions will have to be made about effective shading, high ventilation rates and even night ventilation. In a real-world scenario, this will necessitate high levels of occupant involvement, sophisticated automation and control.

Whether or not occupant wellbeing is improved by the absence of artificial heat (or cooling) is unproven. It could be so if the heating system that it was being compared with was badly designed. On the other hand, a free-running building will normally show much less stability in temperature, both spatially and temporally. However, this may be seen as a positive attribute provided the

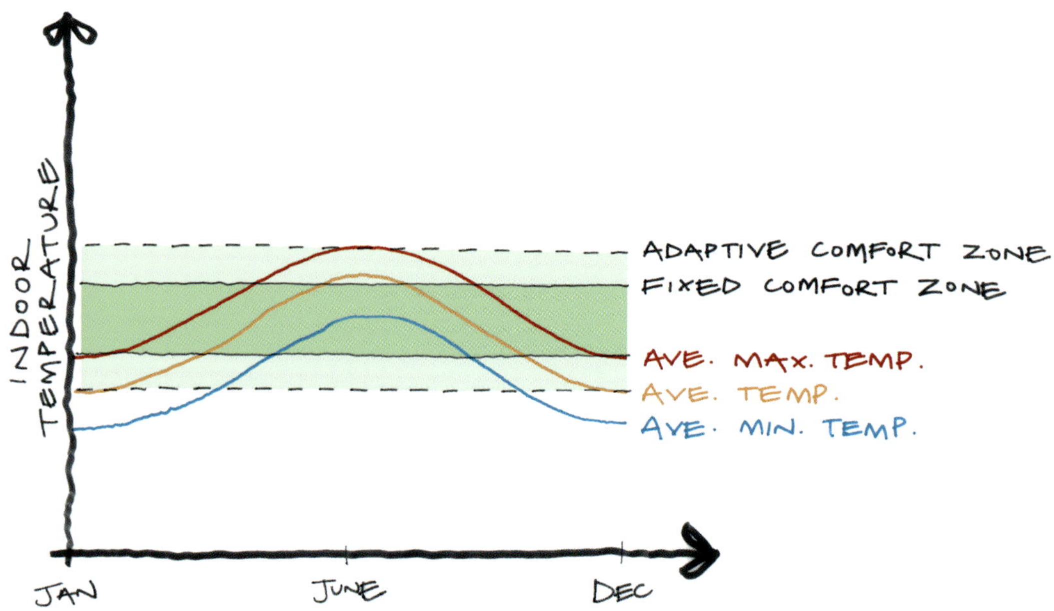

Figure 5.5: Simulated free-running temperature and fixed and adaptive comfort zone.

swings are not too great to be accommodated by the adaptive opportunities available.

Passive solar heating

The matter of temperature swings becomes more challenging if the solar contribution has been maximised by design. In its simplest form, this would be by orientating as much glazing as possible to the south (for the northern hemisphere), and even increasing the glazing beyond that required for sufficient daylight. This approach is known as Direct Solar Gain, as the solar gains occur in the occupied and heated room.

There is an intrinsic comfort problem with direct gain: heat from the solar radiation is made in the occupied room and therefore may affect the occupants directly. Typical solar radiation causes a rise in effective temperature (as felt by a normally dressed person in the direct sun) of between 3 and 6 K. If the room is perceived as cool, this radiant heating would create a pleasant sensation and be welcomed. If, however, the occupant was already warm enough, it would not. Thus, the provision of shady areas in a room, or the easy deployment of shading devices, is an essential part of direct-gain strategy.

Thermal mass plays an important part in this, and has already been discussed in chapter 4's section 'Thermal mass'.

Another approach uses gains made in a secondary or intermediate space such as an unheated conservatory. The transfer of heat from this space to the interior can be controlled by airflow or may be linked to the fresh-air intake, as is referred to as *ventilation preheating*. This approach provides more control and is less likely to lead to discomfort from overheating.

However, with both direct and indirect gain occupant involvement will be needed to work the controls, and there may be difficulties when control is needed when the building is unoccupied. For example, an unexpected sunny afternoon could result in the occupants returning to an overheated living room as there had been no one present to draw the blinds.

A solution to this is to employ automatic controls. However, this carries a significant cost penalty because, apart from the information technology (IT) involved, there is the cost of sensors and actuators – e.g. motors, pumps, hydraulics, etc. to drive the blinds. There is also the need for a smooth transition to occupant control in order to satisfy the adaptive-opportunity principle. This is discussed later in this chapter in 'Heating controls'.

The spatial thermal environment

In most cool climates before the advent of central heating, heat was provided by burning a fuel such as wood or coal in a hearth or stove, in the occupied room. This led to a very particular thermal environment (Figure 5.6).

Due to the inverse square law,[14] the intensity of the radiation – and, therefore, its heating effect on the occupant – falls off very rapidly with distance. Furthermore, especially for an open fire, the requirements for combustion and the exhaustion of smoke sets up a flow of fresh unheated air towards the fire, and leads to an air-temperature gradient and draughts. As described by Witold Rybczynski, in his history of domestic life, *Home*,[15] this harsh thermal condition historically influenced the design of furniture and clothing, and the layout of the room. Clearly, the significant non-uniformity of the environment gave good adaptive opportunities – simply adjusting the distance from your chair to the fire could cover a wide range of comfort temperatures.

Furthermore, the importance of the fireplace as a bringer of comfort meant that it was visually

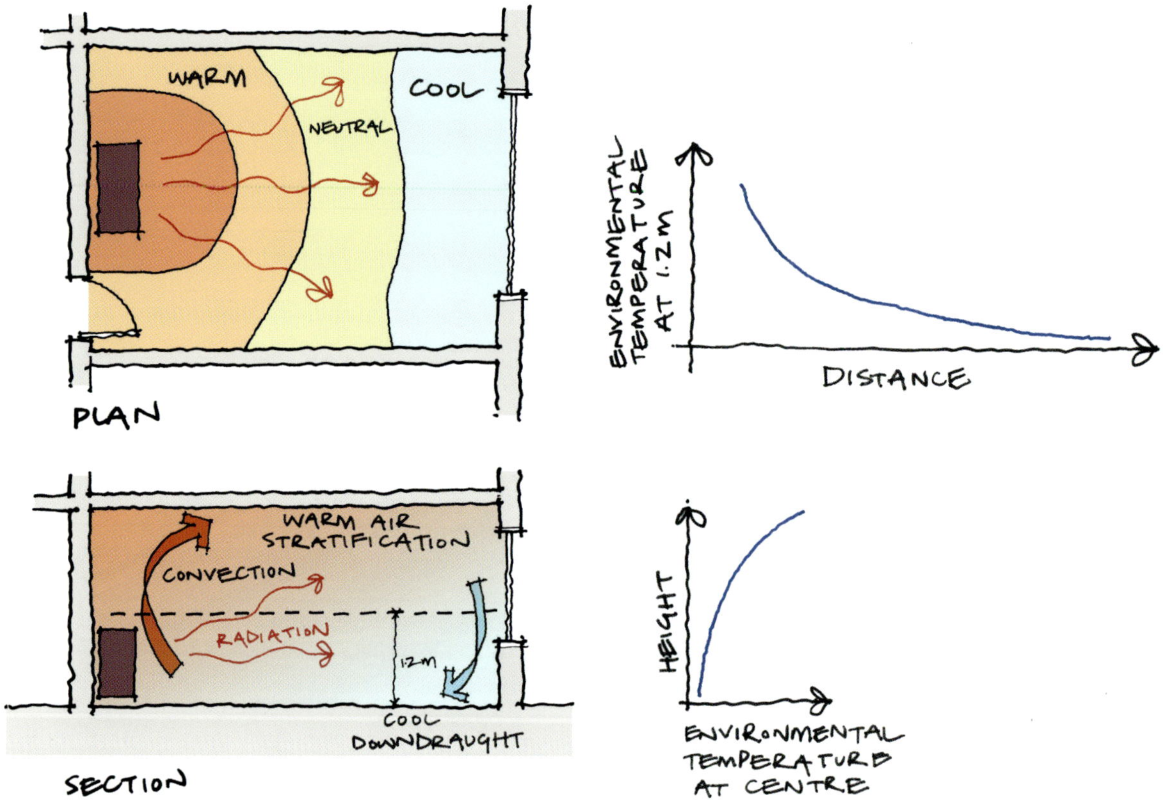

Figure 5.6: The spatial thermal environment of a room heated with a hearth or stove, with vertical and transverse temperature profiles.

celebrated with fancy ironwork, bricks and tiles, and moulded stone or timber; even when unlit in summer, it remained the visual focus of the room. The enclosure of the combustion process – i.e. the development of stoves – reduced wasteful airflow and provided a larger surface for radiation and convective transfer. The result was to reduce the thermal gradient, although it became by no means uniform. Stoves, too, were generally accorded the same visual status as the fireplace and, due to their equally static nature, they had a permanent organising role in the layout of the room.

In contrast, modern central heating has eliminated this organising principle almost completely. The furnace is removed from the space and heat is delivered (usually by hot water) to heat emitters in the room. These are positioned to give as uniform a temperature as possible, which, due to much higher insulation levels, is usually achieved to a considerable extent. Heat emitters – e.g. 'radiators' – are only rarely celebrated visually, and other emitters such as warm-air grilles and underfloor heating are all but invisible.

What impact has this had on comfort and the wellbeing of the user?

Firstly, central heating results in more of a space being reasonably comfortable and therefore useful. As overall space standards have reduced, this is important. However, the problem of feeling too hot or too cold has not been eradicated by modern heating. If a person feels too hot, the room thermostat with its relatively slow response is the only recourse – provided all other occupants are of the same opinion. This contrasts with the kind of minor change of position, posture or orientation that could restore comfort in an older, strongly non-uniform environment.

But there are subtle differences in the thermal environment created by modern emitters that may have impacts on thermal comfort, and even health. Figure 5.7 shows the main characteristic of commonly used emitters: panel radiators, panel convectors, forced convection, heated floors and heated ceilings. Table 5.1 describes their characteristics, advantages and disadvantages.

Heat emitters and safety

Direct contact with the warm surface of a heat emitter such as a conventional radiator can be regarded as beneficial, giving pleasant tactile feedback. However, vulnerable people, the elderly and young children may have impaired responses that could lead to burn injury if the temperature is too high. The accepted surface-temperature limit is 43°C. This closely relates to the water-flow temperature, which is in the control of the boiler thermostat.

However, the heat output is also dependent on water temperature, large heating demands requiring higher temperatures for a given area of emitter. Thus, generous sizing of radiators in order to meet heating demands will ensure lower surface temperatures.

Burn-injury risk can be minimised by the choice of emitter. Low-surface-temperature (LST)

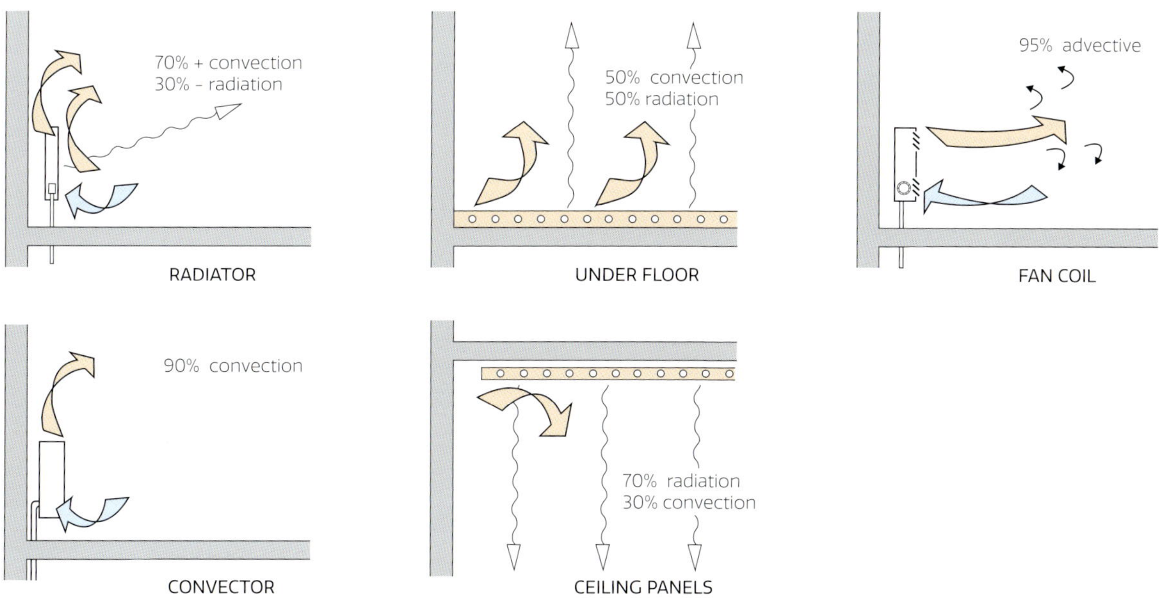

Figure 5.7: Heat emitters and the resulting thermal environment.

panel radiators screen their directly water-heated surfaces with another layer, reducing accessible surface temperature and increasing the convective output. Underfloor heating operates at much lower flow temperatures. Convectors and fan heaters have low surface temperatures provided the airflow is not impeded. A hazard exists of high surface temperatures, even fires, if outlet grilles are obstructed by clothing, curtains, books or toys.

High-level heat emitters such as panels or a heated ceiling remove all risk of skin contact.

Domestic hot water supply and safety

Scalding from domestic hot water (DHW) is also a potential source of injury – particularly to infants, children and the elderly. This can be reduced by limiting the supply temperature to 49°C. However, this temperature is too low to prevent the water-borne bacteria Legionella (responsible for the serious disease of pulmonary legionellosis) surviving in the DHW system. The solution is to heat the water to 60°C and then supply via anti-scald outlets at point of use. These are thermostatically controlled mixer valves as commonly used for showers, but can also be for bath and basin hot-water supply.

Symbolic fire

The disappearance of the fireplace as a spatially and socially orientating element in the home is beginning to be noticed. There has been a resurgence of interest in wood-burning stoves. In some cases, this is in support of a lower-carbon lifestyle despite the challenges associated with emissions of particulate matter and other (non-carbon) polluting gases, but it is often for aesthetic reasons and not the only form of heating. Despite installed central heating, householders are prepared to put up with the inconvenience of wood and other solid fuels, and the resulting mess of ash and release of dust particles.

Few would deny that there is something compelling about a flickering fire. Stove manufacturers go to some length to include glass windows through which the conflagration can be viewed, although this serves no significant thermal function. The action of lighting, stoking, regulating the fire and even the de-ashing is one of those welcome chores. It must be that it puts us in touch with our primeval respect and valuing of fire.

Simulated fires have been around for nearly a century, and predate the widespread use of central heating. In the early 20th century, the UK began a love affair with electricity, largely due to the emergence of artificial lighting and household appliances such as cookers, irons and vacuum cleaners. It is therefore not surprising that electric heating was seen as an appropriate application of the new, clean 'wonder energy': to be modern was to have an electric fire.

These electric fires consisted of uninsulated wire wrapped around a fire-clay tube. When switched on, the element reached a temperature of around 1,200°C and became incandescent. In thermal-environment terms, they became a close replacement for the open fire – i.e. a point radiant source – but without all the disadvantages of dirt, smoke and draughts.

But their functional modernist style did not suit everyone – many missed the glowing coals and flickering flames. So designers produced electric fires with imitation coal or wood, illuminated by a flickering lamp. Half a century later, the full might of IT is brought to bear and now dedicated LCD or plasma screens display hour-long cycles of a burning fire, accompanied by hissing and crackling. Only the smell of smoke is missing.

Is this just a quirky interior-design gimmick, or is it to be taken seriously – is there really a fundamental need to associate warmth and comfort with combustion, and would this

Heat emitter type	Conv/rad split	Comments
Conventional panel radiator	*70/30*	*Leads to vertical temperature gradient during warm-up period. Reduces cold downdraught if under window, but at a cost of increased heat loss. Generous sizing allows lower water temperature, reducing risk of burns and increasing boiler efficiency.*
Underfloor	*50/50*	*More uniform vertical gradient and warmth by conduction with floor contact. Good for young children and sedentary aged. Heat output reduced with carpeted floor unless water temperature increased, reducing efficiency. Cannot meet large local loads – e.g. close to single-glazed windows.*
Fan-coil convector	*95/5*	*Good lateral distribution. Noise generation at higher outputs. Dust disturbance. Can give high output for small size.*
Convector	*90/10*	*Creates strong vertical temperature gradient during warm-up period (high-temperature air floats to ceiling). Poor horizontal distribution. Lightweight ceiling allows rapid warm-up and conversion to downward radiation.*
Ceiling panels	*70/30*	*Good vertical uniformity of air temperature, but higher radiant temperature for horizontal surfaces – e.g. top of occupant's head or back of hands. Good for poorly insulated spaces such as conservatories, allowing lower air temperatures for a given comfort temperature. Most effective as radiators when horizontal (not sloping). Also good for children's floor play on carpet, as radiation quickly warms up carpet surface.*

Table 5.1: Ratio of convective to radiative heat output, and notes on the characteristics of various heat emitters.

increase thermal satisfaction? We think the answer is probably yes.

At the moment, imitation fires are purely visual and auditory and would be placed in a conventionally heated room. But the question then is: should the radiant environment match the visual display? If the display shows a fireplace or stove, should the thermal environment be a radiant field as emitted from a semi-extended point source? This will be a challenge to engineers and architects.

Heating controls

There are two main functions of heating controls: room temperature and programming (timing). The former has a direct impact on occupant comfort; the latter is primarily concerned with energy conservation, and thus not of direct relevance to health and wellbeing. However, since programmed heating in houses is almost universal, and usually involves the same equipment as temperature control – and, furthermore, affects comfort when mis-matched – it will be included here.

Mechanical heating is to be found in almost all housing in central and northern Europe, whilst installed mechanical cooling is relatively rare. Although referring specifically to heating controls, most of the discussion below would equally apply to cooling controls.

Temperature control

Most central-heating systems sense return water temperature (which is dependent on the average house temperature), and thus can offer global temperature control to the whole house. However, due to different temperature requirements in relation to rooms' functions, wall-mounted room thermostats are often installed as well in certain spaces.

The function of a thermostat is to compare room temperature with the set point, and if below the set point to call for heat until the room temperature is a certain value above the set point, and then close off the heat. This causes the room temperature to oscillate around a narrow band of temperature (usually 2–3 degrees), known as the *dead band*.

This function is well known to engineers, but less well understood by occupants. For example: on feeling cold, a typical response of an occupant is to turn the thermostat up to a high value – e.g. 26°C – under the mistaken impression that the room will reach a comfort level more quickly than if they set it to their preferred temperature of 23°C. The result is that the temperature may reach 25°C or 26°C before the occupant feels not comfortable but too hot, and then makes another adjustment downwards. This is clearly unsatisfactory from both a comfort point of view and energy conservation, since the overheating is wasteful.

Furthermore, people fail to understand that unless the room thermostat controls only that room, by use of a motorised valve, the whole building will respond to the adjustment. In other words, its thermal function is no different from the thermostat on the boiler; it merely offers more convenient access.

It is also common to have thermostatic radiator valves (TRVs). These do give local control and are of particular value in rooms where significant heat gains could be made from solar or occupant activities such as cooking. In response to a local rise in temperature above their set value, they will shut down or reduce the heat from that emitter (radiator or other device), resulting in effective local control. Their main disadvantage is that when located at the base of the emitter thermostats often have poor access for the occupant, and again suffer from the *thermostat-misunderstanding*. They are also poorly positioned to sense the representative temperature of the room. Furthermore, they may have no effect in increasing room temperature if the boiler-control thermostat (located in the room or elsewhere) is dominating.

Some of these shortcomings are met using radiator valves with remote sensors and setting units. The unit can be positioned where there is good access and temperature measurement, and the valve unit can control one or more radiators (or emitters) in that room.

A recent development has been the use of wireless connection between the thermostat and boiler, allowing the thermostat unit to be moved to different locations to suit the occupant. The portable units, such as those shown in Figure 5.8, display the current local temperature and the current set point, and offer the facility to change the set point. This may appear to be a step forward, but actually adds to the confusion due to its mobility. For example, if an occupant takes it into a bathroom and sets the temperature to a high value, a person occupying the lounge will suddenly find the temperature increasing. Or, if the unit were placed near an open door or in a downdraught close to a window where locally the temperature is lower than the prevailing house temperature, the whole house temperature will increase unintentionally. On the other hand, if the unit is placed near a local source of heat, such as a cooker, television set or heat emitter, the whole house temperature will be reduced.

In speculating on the ideal heating system, it is interesting to revisit the basics of thermal comfort. A person feels cold discomfort as an early warning that either their core temperature is already below optimum or their skin sensors indicate that the core temperature could become so in the near future. This explains why there is no one ideal temperature for the human being, since it is dependent on the thermal history of the individual – more specifically, the current and previous heat balance between heat loss to the environment and heat gain due to metabolism.

Figure 5.8: Portable wireless thermostats in the sun – a non-ideal location that does not represent the overall condition of a home.

We can reliably define the state of 'being cold' as 'wishing to be warmer', so maybe that is what we should give people. The heat controller could simply have a 'Warmer' (or 'Cooler') button. This would raise the temperature above the mean target value temporarily, and then slowly return to its set point. It would anticipate the gradual recovery of skin and core temperature of the occupant – i.e. acknowledging the dynamic nature of thermal comfort. It has an equivalence, in a primitive setting, of stoking up the fire and throwing on another log.

Interestingly, in the primitive setting, the chilly occupant would almost certainly respond first by placing himself or herself closer to the fire, intuitively making use of the non-uniform radiant field. Sadly, in most modern domestic-heating designs this option has been eliminated, as we have discussed in the previous section. The principle of adaptive control suggests that we should be restoring this option – should be deliberately creating non-uniformity in the thermal environment, as an adaptive opportunity.

Heating programming

The main purpose of a heating programme is to match the heating periods with an intermittent occupancy, and/or heating need. For example, a typical use pattern for a house might be: unoccupied for the daytime on five weekdays, occupied all evenings, and all day and evening at weekends, with no heating required after 12 midnight until 7 am on all seven days. This can be achieved with a programmer, either provided as part of the boiler or as a separate unit.

In most cases, the space heating of a house is on one circuit and therefore controlled by one programmer. This means that bedrooms and day rooms will always be heated together. Another occupancy scenario, for a family with young children, would require dayrooms heated seven days a week and bedrooms heated at night. A single circuit would lead to wasteful heating of bedrooms in the daytime and dayrooms at night. A two-zone heating system could cope with this, by applying a separate programme to each zone.

Does this have any relevance to health and wellbeing? Yes, in the indirect sense of avoiding waste and its impact on the environment. In a more direct sense, it may influence space use and the occupants' perception of space provision. Unfortunately, these two issues are in conflict. One of the long-held beliefs about central heating is that it is beneficial to heat a room *in case* someone wants to use it, or even visit it. Thus, the application of a strict spatial/temporal heating programme could be perceived as a constraint – in particular, a spatial constraint equivalent to having less usable space.

Is there a way forward?

Firstly, we could make programmers much simpler to set, and more attractive for the occupant to interact with and make temporary alterations. This could include a far friendlier interface, and also a return function that recognises an adjustment as temporary. These features are to be found in some IT-based programming and control devices described later.

Secondly, increasing insulation standards to the building envelope has meant that temperature differences between rooms have greatly reduced (Figure 5.9). There would be a strong dis-incentive to leave a day room at 21°C and make unplanned use of a bedroom at a temperature of 10°C, but to occupy the bedroom at 17°C, which on activating a local control would achieve 21°C within 15 minutes, would be quite acceptable.

There is much anecdotal evidence that manipulating the micro-environment to provide comfort is a pleasurable activity. Indeed, it is a fundamental tenet of adaptive theory that (successfully) taking an adaptive action is a pleasurable experience and thus driven by our

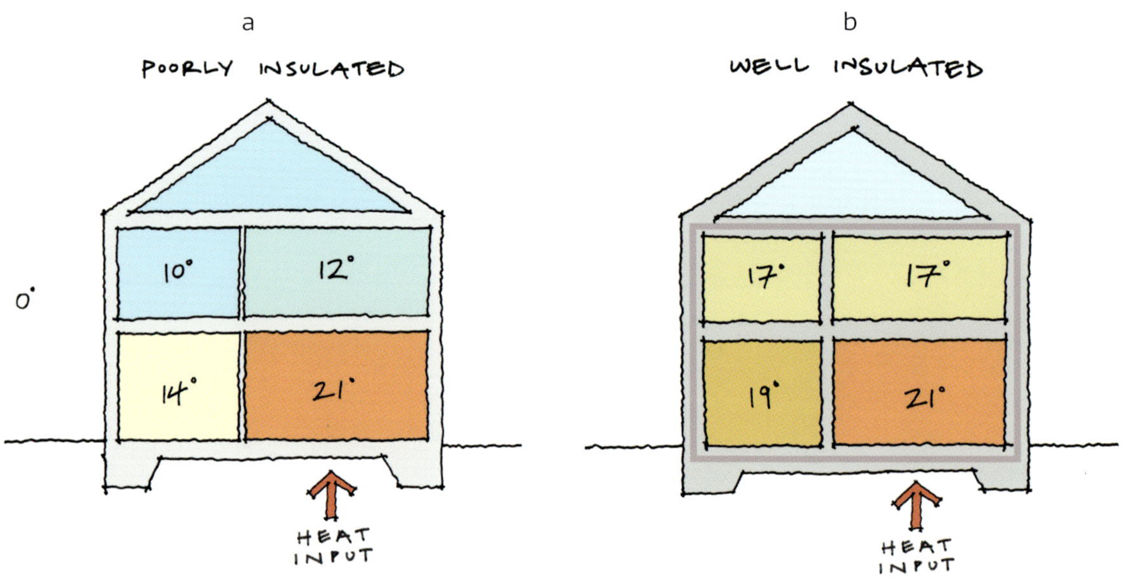

Figure 5.9: Temperature distribution in (a) a poorly insulated house and (b) a well- insulated house.

pleasure-seeking urge. The common perception that in a modern environment 'it shouldn't be necessary' is possibly due either to an individual having wished to make adaptive actions that were prohibited or not available, or to those adaptive opportunities having proved inadequate.

Thus, experiencing cool temperatures on entering an unoccupied room will appear to be 'natural' and the knowledge that a simple action such as pressing a button will within a reasonably short time, render the room comfortable for sedentary occupation will be satisfactory, if not pleasurable.

Smart controls

IT has impacted on controls in four ways:

1. It has allowed complex patterns of demand to be entered and stored easily.
2. It has allowed the use of algorithms that respond intelligently to feedback from signals such as room temperature, outdoor temperature and occupancy.
3. It has made available wireless communication from fixed or mobile interfaces – both locally and via phone and the internet.
4. It can 'learn' the requirements of the user – i.e. their occupancy patterns and temperature preferences.

An example of a smart control package is shown in Figure 5.10.

Figure 5.10: Smart heating controller with optional user access by mobile phone.

It is interesting to examine how compatible this learning behaviour is with adaptive principles. Suppose the occupant has a cold with a slight fever and stays at home for three days to recuperate. On the first day during the normally unoccupied daytime, they have to call for heat and at a higher temperature than normal. However, the controller typically takes a week to learn this pattern, and by then the patient is better and back at work! But before this, their partner may return to an overheated room with the sick person upstairs in bed! Both of these errors could be corrected, by further investment in functions and sensors. This notwithstanding, it seems that the controller is struggling to do something rather badly which the occupant is actually rather good at, and is quite happy to do, provide the response is adequate and the interface is user-friendly.

The design of the interface is a major issue, and poor non-intuitive design often results in non-use, misuse or the fear of mis-use – as illustrated in Figure 5.11.

Caretaker controls

We have invented this term for controls that allow maximum human intervention but always return systems to an energy-minimising state whilst still providing a base level of comfort. For example, the 'Warmer' setting on the thermostat, referred to earlier, would revert back to the long-term value after a period of, say, 30 minutes.

Already, most central-heating controllers have a manual button that simply moves the system

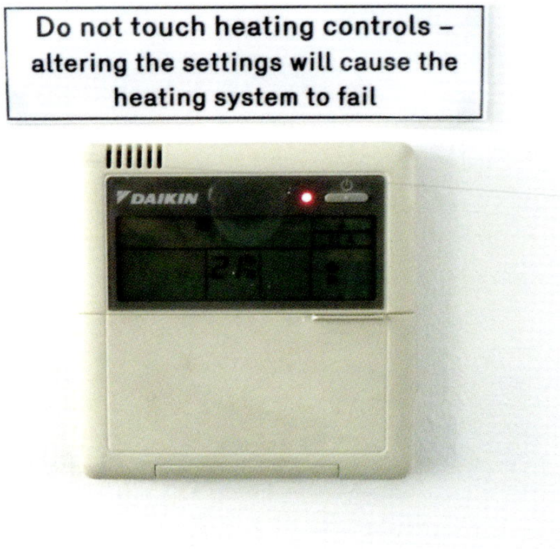

Figure 5.11: Notice seen in an architectural-school classroom. So much for adaptive opportunity!

to the next state in the programme (either on or off) and thereafter rejoins the set programme. Thus, on an unusually early arrival home, the occupant can move the 'On' period forward and activate heating without altering the permanent programme. The same function could be used on leaving the house unoccupied. The manual button would then switch off the heating early, but the programme would switch on the heating at the programmed *switch-on* time.

However, the controller is often in a poorly accessible place – in a utility room or under the stairs. If a wireless control button were to be placed next to the main-entrance door, rather like a light switch, it would be much more likely to be used. This suggests that it is the combination of anticipating human interaction, applying IT intelligence to 'caretake' the interaction, and thoughtful ergonomic and intuitive design of the control interface that is the way forward.

Key points

- Adaptive opportunities such as openable windows, shading controls, lighting dimmers and manual-heating controls
  - enable an occupant to increase the 'depth' of the comfort zone by adaptive behaviour
  - should be provided to enable such adaptive behaviour.
- Lack of adaptive opportunity results in stress over small departures from neutrality.
- Occupants are less stressed in buildings that they perceive have adaptive opportunity, even if they don't use it.
- Extreme indoor temperatures
  - can have serious effects on the physical and mental health of occupants – especially vulnerable people.
  - are typically caused by a combination of high solar gains, poor insulation and poor ventilation. The solar gains may be via unshaded windows or, on older buildings, through poorly insulated fabric – especially roofs.
- Extreme indoor temperatures can have serious effects on the physical and mental health of occupants – especially vulnerable people.
- Overheating is typically caused by a combination of high solar gains, poor insulation and poor ventilation. The solar gains may be via unshaded windows or, on older buildings, through poorly insulated fabric – especially roofs.
- Well-insulated buildings may overheat due to unshaded windows, unavailability of thermal mass and poor natural ventilation.
- Underheating

  - is typically caused by a combination of inadequate mechanical heating, poor insulation and possibly over-ventilation due to a leaky envelope.
  - is made more likely by fuel poverty, to the extent that is can pose a danger to health.
- The free-running building should deliver non-extreme conditions that can then provide comfort by means of plausible adaptive behaviour.
- Passive solar-heating design must make a compromise between gathering solar heat, to offset mechanical heating, and occupant comfort.
- Different heat emitters create thermal environments that are spatially different and should relate to the anticipated room function.
- Heating controls should respond to and affect conditions in the occupied space, be simple to use and have a fast response.
- Heating programmers (time clocks) can save energy, leaving resources available for sufficient heating during occupation.
- *Caretaker controls* offer a 'Warmer'/'Cooler' function that returns to a set point after a delay.

Ventilation and indoor air quality

Ventilation is the replacement of the air inside a building with air from outside. The term *ventilation* is usually applied when the air is admitted and vented through controllable openings. However, building construction is rarely completely airtight and therefore some air is exchanged through cracks and unintentional gaps. This is usually referred to as *infiltration*. Its effect is indistinguishable from and additive to that of ventilation. Both are driven by thermal buoyancy and wind pressure: this is usually referred to as *natural ventilation* (NV). Ventilation can be driven by fans, usually acting in ducts: this is referred to as *mechanical ventilation* (MV).

Ventilation rate is usually quoted in *air changes per hour* (ac/h). This means the number of times that a volume of air equal to that of the building (or room) has entered (and left) the space in one hour. It is related to the volume flow rate **v** in m^3/sec by:

$$v = V \cdot n / 3{,}600$$

where **V** is the volume of the space and **n** is the number of air changes per hour (ac/h).

The driving forces of natural ventilation

There are two driving forces of natural ventilation:

1 thermal buoyancy
2 wind pressure.

To move air through the resistance of an opening or duct requires a pressure difference. Thermal buoyancy (or the stack effect) occurs where the difference of density (due to a difference of temperature of a mass of air compared to the surrounding air) creates a pressure difference, as shown in Figure 5.12. Wind pressure is generated when moving air impinges on an obstruction such as a building and has to alter its direction of flow, as in Figure 5.13 The ventilation rate is also dependent on the opening size.

In reality, both forces are highly variable – and this leads to particular problems, which we will deal with later.

In general, thermal buoyancy and wind pressure are both operating at once. The two mechanisms are not purely additive; at low wind speeds, air flows tend to be dominated by buoyancy, and vice versa – as shown in Figure 5.14. The dashed line shows wind effect only; in this example, at wind speeds of less than 2m/s it can be seen that thermal buoyancy driven by a 4°C temperature difference maintains natural ventilation above 0.8 ac/h.

Wind-generated pressure patterns across building surfaces are complex, as shown previously in Figure 4.8. They also vary in time due to turbulence. These phenomena increase the possibilities of wind-driven ventilation since air will flow inwards from any opening on the surface,

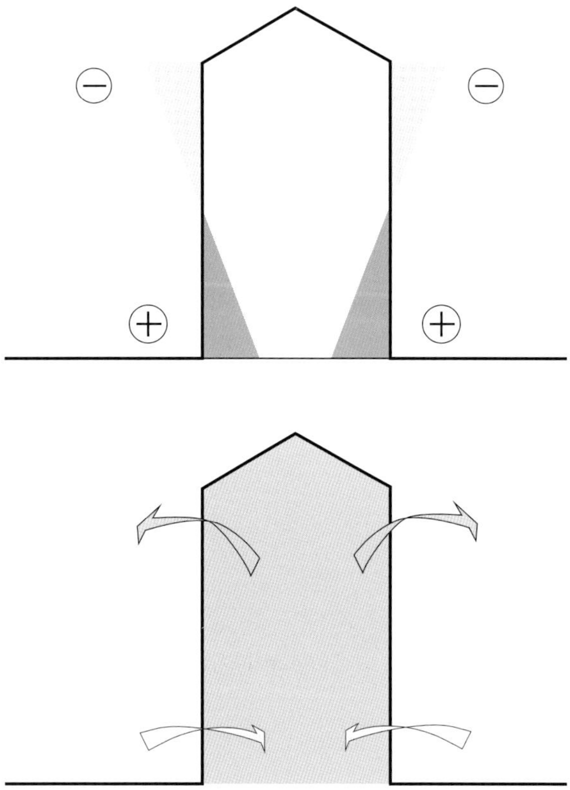

Figure 5.12: The thermal buoyancy of the warmer column of air inside the building generates a pressure difference, driving air in at the ground floor and out at the first floor.

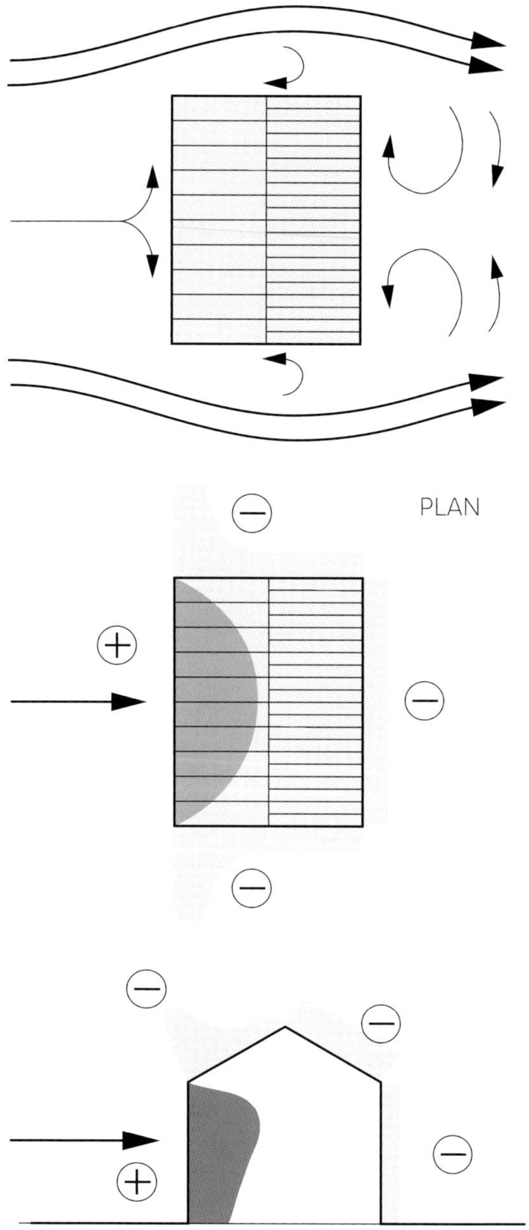

Figure 5.13: The change of direction of the air as it passes over and round the building creates positive and negative pressures over the building envelope.

and out at another where the pressure is less but not necessarily negative (see Figure 5.15). This supports a general principle that several small openings, well distributed over building surfaces, will be better than few larger openings. This also applies to the vertical pressure differences due to buoyancy.

The complex pressure distribution can be exploited in another way. Suppose residential buildings are located downwind of a noisy and polluting highway – how can wind-driven ventilation be used to deliver clean air noise-free?

By making the windward wall closed and airtight, and by connecting to the area of strong negative pressure on the roof via a plenum and stack, cleaner air will be drawn from the side away from the road, from the area of weaker negative pressure (Figure 5.16). In the absence of wind, when driven by stack effect, the same flow pattern will prevail.

A similar approach was proposed by Nick Baker[16] to ensure that ventilation air was collected via a south-facing conservatory (sunspace) as a solar strategy referred to as *solar ventilation preheating*.

The driving forces described above are also highly variable over time. This variability is caused by fluctuations in wind speed and temperature differences that bear no correlation to the ventilation demand. The result is that even if the *average* ventilation rate is appropriate there will be frequent excursions into over-ventilation and under-ventilation – the latter resulting in poor indoor air quality (IAQ). Ensuring that the *minimum* ventilation is always sufficient will result in the *average* being much too high, and hence a waste of energy.

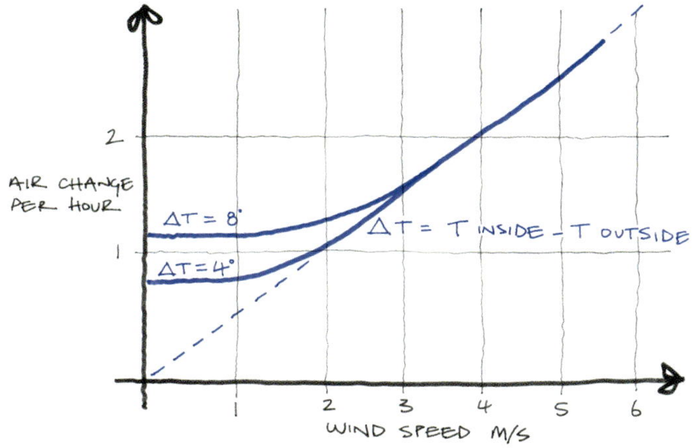

Figure 5.14: Ventilation rate as a function of temperature difference and wind speed.

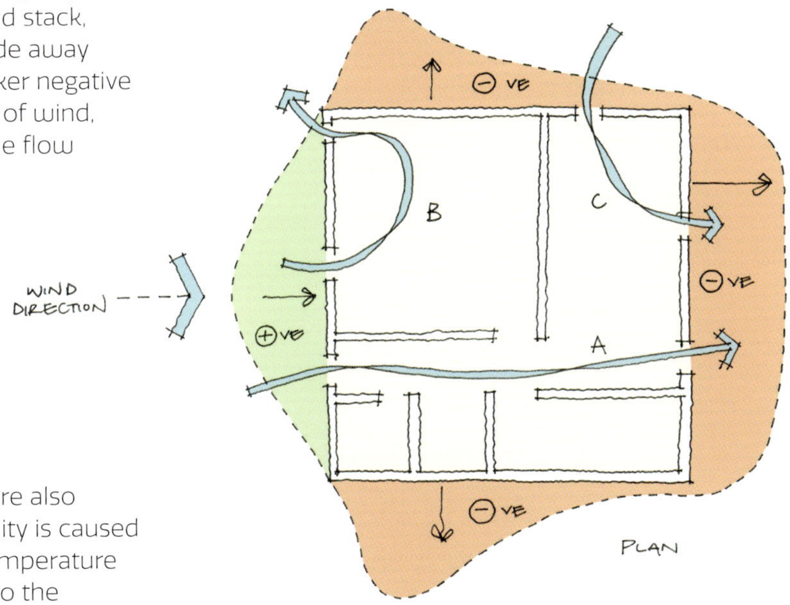

Figure 5.15: Wind pressure on the surface of a building. Ventilation air will flow through the building whenever two openings are at different pressure. By controlling the distribution of openings and exploiting the pressure distribution, the point of ingress of fresh air can be controlled. Path A – positive to negative, path B – positive to less positive, path C – negative to more negative.

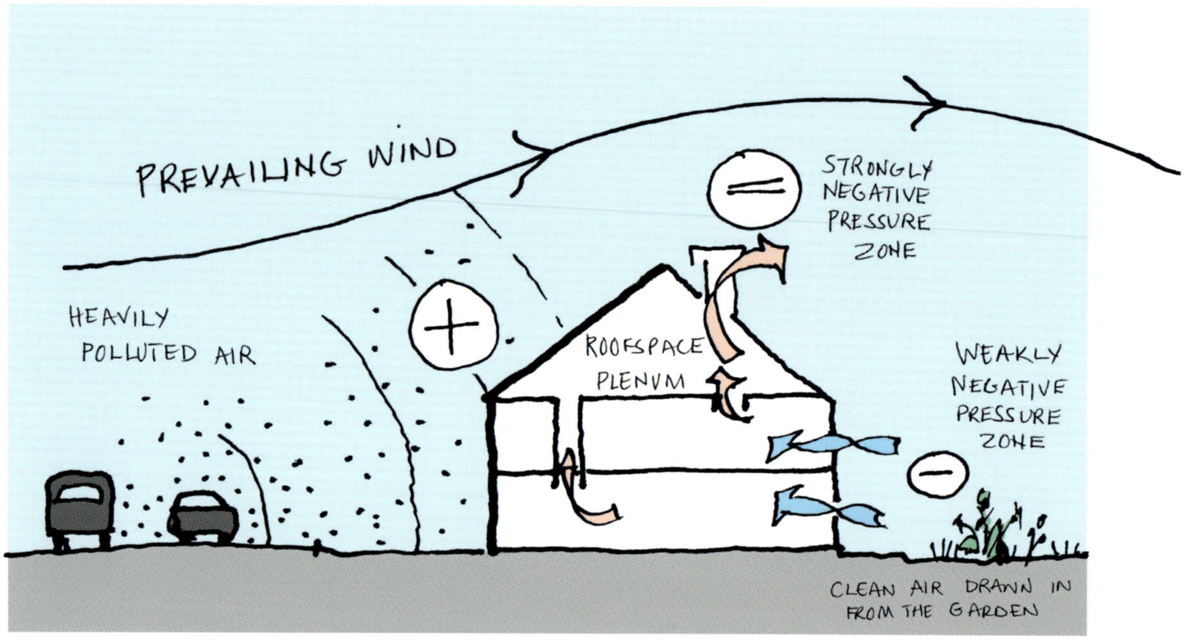

Figure 5.16: Internal pressure is lowered by connecting to a strongly low-pressure area on the roof, causing non-polluted air to be drawn in from the garden.

Technology has a partial answer to this problem: self-balancing ventilation grilles. These devices, fitted into an aperture in the external wall, adjust under wind pressure to give a constant flow rate – i.e. at low wind speeds they offer low resistance, and vice versa. This function could be provided with greater precision using a centralised control system. Sensors measure wind speed and an indicator of IAQ such as CO_2 level, and set the ventilation with motorised actuators.

But before rushing to highly technical solutions, we must not forget that humans are quite adept at controlling their environment. We are good at detecting when a room is too hot, stuffy or smelly and quite capable of taking appropriate action if available. Studies have shown that window-opening habits respond to physical parameters such as temperature, humidity and room use.

Thus, if we can provide a baseline minimum ventilation level, intermittent mechanical extraction in areas of high pollution and well-designed opening windows then, if external conditions are clean and quiet, this is the preferred option.

The three regimes of ventilation

We can identify three distinct regimes of ventilation, corresponding to three 'bands' of ventilation rate and three distinct sets of functions:

1. The provision of a minimum air quality (dilution and removal of pollutants)
2. The removal of unwanted heat gains (to avoid overheating)
3. The provision of air movement for direct physiological cooling.

Regime 1

When ambient temperatures are low, only the first regime is required – i.e. maintenance of air quality – and the relatively low rates can easily be provided by wind or stack effect at normal domestic occupancy densities.

Occupant metabolism generates CO_2 and water vapour, and both are pollutants in the sense that they have harmful effects when highly concentrated. Metabolism also requires oxygen. Regime 1 achieves both functions – i.e. diluting pollutants and providing sufficient oxygen.

If pollutants are being generated – in association with occupancy activity such as cooking, clothes washing or cleaning, which may include water vapour – higher rates of ventilation may be required, as discussed in the following two sections.

Regime 2

In residential buildings, the second regime (removal of heat) is usually only required at higher ambient temperatures. Because temperature differences are smaller, high ventilation rates are needed to remove significant amounts of heat – and thus, large openings are required. The stack effect may be barely adequate and, in the absence of wind, mechanical ventilation may be necessary.[17] Regime 2 is inclusive of Regime 1 – i.e. the non-thermal air-quality criteria will be easily surpassed.

Regime 3

Normally, only wind pressure will be able to generate useful air movement for physiological cooling.

For the second and third regimes, high air-change rates are required – necessitating large openable areas. These are normally provided by windows, thus combining the three functions of daylight, view and fresh air into one element and reinforcing intuitive control by occupants. Typical air speeds that cause useful convective cooling without inconvenience are around 0.5 to 1 metre per second. This velocity over an open area of 1 m^2 would generate an air-change rate of about 15–20 ac/h in a domestic-sized room. Thus, regimes 1 and 2 are included in Regime 3.

However, there may be some circumstances in which, due to ambient noise or pollution, opaque ventilation components (flaps, dampers, etc.) incorporating sound attenuation and/or filtration are required. In terms of ambient noise control, the design of a window can have considerable influence, as illustrated in Figure 4.7, and this should be considered when specifying retrofit windows.

Ceiling and desk fans are the mechanical alternative to providing air movement to enhance convective and evaporative cooling, but in this case the functions 1 and 2 will not be met since there is no exchange of internal and external air. Indeed, fans can cause a reduction in air quality due to the re-suspension of dust or, in the case of ceiling fans, the redistribution of stratified vitiated air, thereby spreading airborne contaminants and infections.

Indoor Air Quality

IAQ is a term used to describe the degree to which indoor air is ideal for supporting human life and free from pollutants that have either a direct or indirect effect on health and comfort. There has been intense interest worldwide in this issue over the last three decades. The WHO estimates that 4.3 million people die each year from indoor air pollution. Most of this occurs in poorer countries where carbon monoxide and particulate carbon from inefficient cooking appliances are the main culprits. However, with affluence comes hazards from contents – carpets, paints, plastic laminates and adhesives are all sources of emissions toxic to humans. Furthermore, the ubiquitous motorcar brings pollution into the heart of almost all settlements, degrading the air even before it enters the home.

IAQ is too large a large subject to deal with comprehensively here. Rather, we present the underlying principles and recommended strategies. For further information, the reader is referred to specialist literature in the 'Further Reading' and 'Bibliography' sections of this book.

'Pure' air is a mixture of about one fifth (1/5) oxygen and four fifths (4/5) nitrogen, with a small fraction of CO_2 (0.04%), a variable amount of water vapour (H_2O) (around 1%) and a few other trace gases.

Air inside a building departs from this ideal combination for a number of reasons. Firstly, in a residential situation, the occupants produce CO_2 and H_2O as a product of respiration – this increases the normal concentration. Secondly, building contents – e.g. building materials, finishes and furnishings – may emit gases of varying levels of toxicity over a period of time. Thirdly, processes and activities carried out in the building may cause emissions, such as cooking, cleaning, use of adhesives and the production of particulates (dust). Finally, the air quality inside the building will be influenced by that outside it. The main culprit here is transportation and the internal-combustion engine emitting hydrocarbons, nitrous oxides and particulates.

In residential buildings, most problems from internal sources will occur during the heating (or cooling) season when the ventilation rate is minimised in order to save energy. If the building ventilation is in regimes 2 or 3, then due to the high ventilation rate it is unlikely that the air quality will be poor unless the outdoor air is polluted.

There are three broad approaches to maintaining adequate IAQ:

1. Reduce or eliminate the input of the pollutant.
2. Dilute the polluted indoor air with clean air, to bring the concentration of the pollutant down to acceptable levels.
3. Recirculate the air mechanically and extract the pollutant by filtration, chemical or electrical means. This function is usually present in air-conditioning systems.

CO_2 and water vapour are unavoidable pollutants resulting from occupation and thus, for residential buildings without air conditioning, can only be dealt with by the second strategy – dilution. This is normally achieved by natural ventilation (NV).

Pollution sources – occupants, contents, activities, outdoor air

Occupants

Minimum fresh-air supply rates per person have been established – a typical figure is 8–10 litres/sec per person. This is considerably more than is required for the supply of oxygen and is actually set by the need to dilute CO_2, H_2O and body odours to an acceptable level. High CO_2 levels lead to drowsiness, high levels of H_2O (usually indicated by high relative humidity – see 'Moisture control and condensation' in chapter 4) could lead to skin-wettedness discomfort and high levels of body odour are considered unpleasant. Indeed, on the last-named issue, although body odour is not in itself dangerous we can consider it to be an 'evolved proxy' that indicates risk from other harmful agents that cannot be sensed – in particular, CO_2. It may also indicate the increased risk of transmission of infectious diseases.

About one quarter of UK homes are also home to a dog or cat. Particles of fur and skin, (often referred to as *pet dander*), are a cause of allergic reaction in some people. The extent of the problem may be dependent on the actual variety of dog or cat, its husbandry – grooming, etc. – and also the choice of materials and finishes in rooms that it occupies.

Building contents, construction materials and finishes

Construction and room contents contribute to indoor pollution in two ways:

1. as new materials, they may generate gaseous pollutants that are remnants of their manufacture
2. contents may contribute to indoor pollution simply by acting as reservoirs for pollutants

The first instance is particularly common with fresh paint, new carpets and furnishings, and plastic laminates. The process, known as off-gassing, will generally reduce over time. Most of these chemicals are *volatile organic compounds* (VOCs), and manufacturers are obliged to give information and ratings about them – as they are for most materials found in buildings. VOCs may not be highly toxic but there is evidence that long-term exposure to them can be harmful to health. A particular danger lies in emissions that do not cause perceptible odours.

There is a common misconception that 'natural materials' are free from harmful VOCs. This is not entirely true as, for example, some timbers have volatile compounds – hence their smell. However, more importantly, natural materials may have been treated with preservatives, fungicides, insecticides or polishes – all of which are likely to have VOC emissions, at least in their early life. The emission is often from the solvent that delivers the permanent component of the treatment or process. Increasingly, water-based treatments are greatly reducing VOC emissions.

In the second case, the effect of a short-term pollution input – such as the presence of an animal or smoker, or VOCs from some cleaning or DIY activity – can be extended over a longer period. Carpets, soft furnishings and bedding are notorious for harbouring dust particles and providing a home for allergen-producing organisms such as the dust mite.

An obvious design response is to avoid materials that are in themselves generators of pollutants or can act as pollutant reservoirs. In the specification of construction materials, steer clear of those associated with chemical off-gassing or biological pollution within the home, and specify non- or low-emitting adhesives and alternatives. Examples to support such specification can be found in LEED,[18] the WELL Building Standard (both already encountered in chapter 1; the latter provides additional advice on limiting toxic materials in construction)[19] and the International Living Future Institute (which provides a 'Red List' of chemicals in construction that are to be avoided).[20] Another approach is to choose materials with their cleaning needs in mind. However, much of the contents of an occupied room comprise furniture, clothing and other objects that are controlled by the occupant rather than the designer.

Indoor activities

The number one villain here is undoubtedly smoking. There has been much publicity about the hazards of smoking, and of passive smoking, and fortunately it is becoming common for smokers to refrain from lighting up indoors, even in their own homes. Smoking generates carbon particulates and carbon monoxide as well as nicotine – all of which are harmful.

However, for many we could say that smoking is a fact of life despite the risks. Whilst we are not advocating smokers having to sit with their heads in extractor hoods, it might be possible to introduce a zone in a room where local mechanical extraction could provide higher rates of air change and prevent flow into other parts of the space. A century ago, similarly to moving your chair (seen above in 'The spatial thermal environment'), this could have been achieved by simply sitting close to an open fireplace.

Whilst worldwide, cooking is identified as a serious source of indoor pollution, in developed countries the use of gas and electricity in modern enclosed cookers, together with mechanical extraction at the point of pollution, has greatly reduced the hazard. Furthermore, as discussed in 'Eating', (chapter 2) preparing fresh meals at home has numerous health benefits (Figure 5.17). As with smoking, the main hazardous agents are particulate carbon and carbon monoxide – these are associated with burning, which apart from accidental events, is often part of the cooking process.

More recently, attention has shifted to the effects of deep-frying. Hot cooking oils emit polycyclic aromatic hydrocarbons (PACs), which have been shown to have carcinogenic properties.[21]

In a domestic situation (unlike a commercial kitchen), exposure is usually relatively brief. However, it is within the capabilities of ventilation technology to provide high-performance extraction at source. Improved design of extraction hoods that ventilate to the outside (even enclosures like fume cupboards, as used in laboratories) should be adopted – particularly in open-plan domestic situations. This will not only bring health benefits but also protect decorations and control cooking smells.

Cooking can generate large quantities of water vapour, which, although not toxic in itself, if allowed to escape can cause surface condensation and mould growth (Figure 5.17). Spores produced by the moulds are known to be dangerous allergens, often blamed for asthmatic attacks.

Hobby activities such as model building and painting can be a source of VOCs and particulates generated, for example, by sanding operations on materials such as wood and medium density fibreboard (MDF). Similar pollutants can be produced during DIY activities, by paints and adhesives; these may be generated in large quantities but for relatively short periods. Proper protection for the people involved should be the norm – e.g. respirator masks and local extraction – but in the case of household DIY, avoidance, where possible, of materials with pollution potential is a good strategy.

Figure 5.17: Despite the pleasures and health benefits of preparing fresh food, cooking can contribute to indoor pollutants, water vapour and heat if not effectively controlled with local mechanical ventilation extract.

Ventilated cupboards

Ironically, washing and cleaning are sources of indoor pollution. In the case of clothes washing, possibly harmful pollutants are associated not only with washing liquids and detergents but also fabric softeners and conditioners.[22] Exposure may not be limited to the duration of the washing activity but could extend into the airing process and storage, usually taking place in airing cupboards and wardrobes ventilated to the room – often bedrooms. Using airing as a means of drying will contribute to raising the relative humidity of the indoor air, increasing the risk of mould growth.

The first step to mitigate this is to encourage the use of outdoor clothes drying by providing facilities such as clothes lines and external covered drying spaces. The second step is to ensure that cupboards for clothes and linen airing and storage are sealed to the room with well-fitting doors and have ventilation to outside. This is relatively simple if the cupboard is on the outside envelope: a small trickle ventilator top and bottom. If not on the perimeter, a passive-extract chimney could be considered but this will require an inlet; if drawn from the room, there is a risk that backflow could defeat the object and deliver polluted air into the room.

Many household cleaning products are also hazardous. Not only can they contribute to pollution of indoor air but they are often toxic themselves, posing a risk from accidental drinking by young children or even disorientated or dementia-suffering adults. This suggests that a 'safe' should be considered for all dangerous household substances, providing physical security. Since its contents may themselves be a source of air pollution, this secure cupboard should also be well sealed from room air and ventilated to the outside, as described above.

Finally, it is ironic to report that there is evidence that intentional 'pollutants' such as air fresheners and perfumes may in fact be hazardous.[23] Symptoms associated with exposure to realistic levels of air fresheners include migraine, headaches, asthma attacks, breathing difficulties and mucosal symptoms. Clearly air-freshener use cannot be controlled directly by design, but the provision of means to ventilate well and to control internal pollution – e.g. well-designed openable windows and appropriate use of mechanical extract, together with inherently low-emission materials – will reduce the incentive to use air fresheners.

Vacuum cleaning

Mechanical cleaning activities disturb dust. In particular, portable vacuum cleaners have the potential to redistribute dust into indoor air and cause serious pollution. There are two approaches to reducing this hazard. A high-specification filter, capable of collecting particles down to 0.25 microns in size, can be obtained for some devices. This should ensure that the recirculated air exhausted from the machine is free from harmful particulate matter such as PM2.5,[24] mould spores and dust-mite pellets. However, it relies on regularly changing the filter and will still allow gaseous pollutants and odours that have been released by the cleaning process.

A better approach is to use a centralised vacuuming system. This consists of a central motor and extractor fan that exhausts to the outside. Ducts with one or more connections in every room link the unit to the vacuuming tool. Some central systems lack an external exhaust, and are promoted for improved convenience only. In principle, it should be possible to provide external exhaust connections for portable vacuum cleaners; however, at the moment there seems to be no such system on the market. Centralised vacuum systems with external exhausts are a good services-design response, and could make a real contribution to improving IAQ.

Ambient air pollution

Outdoor air can contain pollutants such as nitric oxide (NO), other oxides of nitrogen (NO_x), carbon monoxide (CO), VOCs and particulate carbon down to 0.25 microns. All of these are associated with road and air transport, and may also be caused by combustion for building heating. Factors and planning issues relating to outdoor air quality, and its distribution at the urban scale, have been dealt with in 'Urban wind patterns' in chapter 3.

Indoor air quality is limited by the quality of outdoor air from which ventilation air is drawn, unless some kind of filtration process is applied. However, pollution distribution is not uniform around buildings. Where the pollution source is local – e.g. a busy nearby road – pollution levels will be much higher on the road side of a row of houses than on the garden side. The decision for the living area to have openable windows to the garden and smaller sealed windows onto the busy road may have already been driven by noise, view and privacy considerations, and will also result in lower pollution levels in the incoming air. This simple strategy is not without problems, however, since during periods of wind from the direction of the pollution source, airflow of the cleaner air from the downwind side may be reduced or even reversed.[25] This effect can be minimised by careful distribution of openings, as discussed in 'Natural ventilation' and 'The driving forces of natural ventilation' in chapters 4 and 5 respectively.

Installed outdoor-air filtration is another option. It is technically possible to filter out particulates and some gaseous pollutants.[26] Manufacturers claim that particulates down to 0.25 microns, including mould spores and bacteria, can be eliminated. Road-traffic pollution can be reduced, with removal of NO_x and particulates, but some gaseous pollutants – e.g. CO – cannot be removed. Filters can be installed in the inlet to whole-house air-conditioning systems or mechanical-ventilation systems. As with all filtration plant, maintenance is essential in order to ensure filtration performance – and this can be a problem in a domestic situation.

For natural ventilation, filtration is virtually out of the question due to the relatively low pressures available. Clearly, this means that openable windows are not suitable for this strategy and that the psychological benefits of the openable window would be lost. Mechanical solutions also require ducting and room grilles and, together with electrical controls, represent a general escalation of technical investment.

Portable air cleaners are more common than installed intake systems. These consist of freestanding units that recirculate room air via chemical and mechanical filters. Since they are located in the occupied space, they act on pollutants generated indoors and brought in from outdoors. However, their performance will be strongly influenced by the maintenance of the filters, especially in conditions of high pollution.

Radon and other ground gasses

Radon gas is a radioactive daughter product of radium, uranium and thorium, and is found in the soil in various well-defined geographical areas.[27] It is a clear, odourless heavy gas and may be found in ground floors and basements of buildings erected in risk areas. Radon and its decay products can attach to dust particles and be ingested into the lungs, causing lung cancer. Mitigation includes the isolation of internal spaces from the ground using impermeable membranes. It is regarded as a major health hazard and there is much specialized literature available on its effects and mitigation. If a site is located in a radon-rich area, it is recommended that specialist advice is sought.[28 29]

Other ground gases are also hazardous. Methane and CO_2 are found naturally in soils containing large amounts of decaying organic materials, such as marshes, and in larger quantities in made-up ground over landfill sites. In some cases, landfill gases are present in such quantity that they warrant extraction and use as a combustible gas. Again, for mitigation, specialist advice should be sought.[30]

Moulds, fungi and bacterial growth

Mould and fungi reproduce through airborne spores, tiny microbial particles. They are known to cause respiratory symptoms, allergies and asthma as well as disturbance of the immune system. Their actual quantitative impact on morbidity and mortality is still uncertain, partly because the prevalence of mould and damp in homes is associated with other risk factors such as poor diet and unhealthy lifestyle. However, it is undeniable that the sight and smell of mould alone is likely to have a negative effect on wellbeing.

The WHO identifies dampness in buildings as the main cause and an essential condition of the problem: 'Excess moisture on almost all indoor materials leads to growth of microbes, such as mould, fungi and bacteria, which subsequently emit spores, cells, fragments and volatile organic compounds into indoor air'.[31] A review of surveys in the US, Canada and Europe reveals that dampness was found in 20% of buildings surveyed.

The four common sources of dampness in buildings are:

1. water penetration from outside the envelope due to porous material, cracks or poor drainage
2. water migration from the damp ground by capillarity action through porous building materials
3. water leakage from plumbing and drainage
4. water produced inside the building by the occupants and their activities.

In new buildings, prevention of the first three sources is covered by the Building Regulations, and chronic dampness conditions from these causes are rare in recently built houses. However, they may still be found in the older building stock – in particular, in solid-walled masonry buildings and those that have had solid concrete floors without damp-proof membranes. A number of techniques and products can be applied retrospectively to mitigate these effects, such as injected damp-proof courses and two-pot liquid-applied resin membranes.

The fourth source of dampness continues to be an issue in new buildings, since it is primarily due to the occupants – that is, the balance between moisture production by cooking, clothes washing and drying, house cleaning and ventilation. Modern standards of airtightness, mean that uncontrolled infiltration is no longer adequate as a failsafe. The role of ventilation to reduce indoor humidity, and hence condensation risk, has already been discussed in 'Moisture control and condensation' in chapter 4. Here, we draw attention to the fact that even without surface condensation being present, at high humidities mould can grow on many building materials and finishes.

Humidity and mould

It is generally agreed that mould growth requires prolonged humid conditions of an average RH greater than 70%. Materials based on organic substances such as wood, cotton and flax – all containing cellulose – are especially prone to mould growth. Their hydrophilic (moisture-loving) properties mean that their exposure does not have to be continuous; a time-average in excess of 70% RH will be sufficient.

Without a quantitative hygro-thermal model of a house and its moisture inputs, it would be difficult to predict precisely the humidity conditions and the risk of mould growth. It is usually sufficient to reduce the risk by:

- minimising the input of moisture by extraction at source
- providing sufficient whole-house ventilation to maintain air quality as demanded by CO_2 and odour criteria (see above).

Surface temperature

The temperature of interior room surfaces may be significantly lower than the main body of air in them. This will be the case for the external walls in a heated building; well-insulated surfaces will be close to average room temperature but surfaces such as windows, even if double-glazed, are likely to be several degrees below it. Poorly insulated surfaces or areas of thermal bridging may also be at significantly lower temperatures than that of the room air. Thermal bridges occur where there is an incomplete insulating envelope, usually due to the intrusion of a dense non-insulating structural element such as a floor slab or column.

Moisture from the air will diffuse into the boundary layer of cooler air adjacent to a cool room surface, and will increase its relative humidity (Figure 4.16 shows how warmer air can carry more water vapour). The humidity may become sufficient to support mould growth on suitable materials, although average room air is dryer than this. This points to the importance of good building-fabric insulation and the avoidance of cold bridges.

There may be particular difficulties when retrofitting insulation. If insulation is applied to internal walls, problems arise where floor slabs meet the structural walls. This can be exacerbated when the floor is carpeted; due to its insulating properties, the carpet in effect causes local interstitial condensation (see 'Moisture control and condensation' in chapter 4). Even when insulation is placed outside, problems may be encountered where balconies are cantilevered from the walls by concrete or steel elements. Lowered surface temperatures might also be experienced in areas adjacent to static air, such as in corners and behind furniture.

Mechanical ventilation

Various configurations of mechanical ventilation are shown in Figure 5.18, almost all of which can be found in residential buildings. Details of mechanical-ventilation design are not given here, and the reader is referred to specialist texts found in the 'Further Reading' section. Here, we give a broad overview together with implications for health and wellbeing.

Local extract

Extractor fans are the most common form of mechanical ventilation. They are usually installed

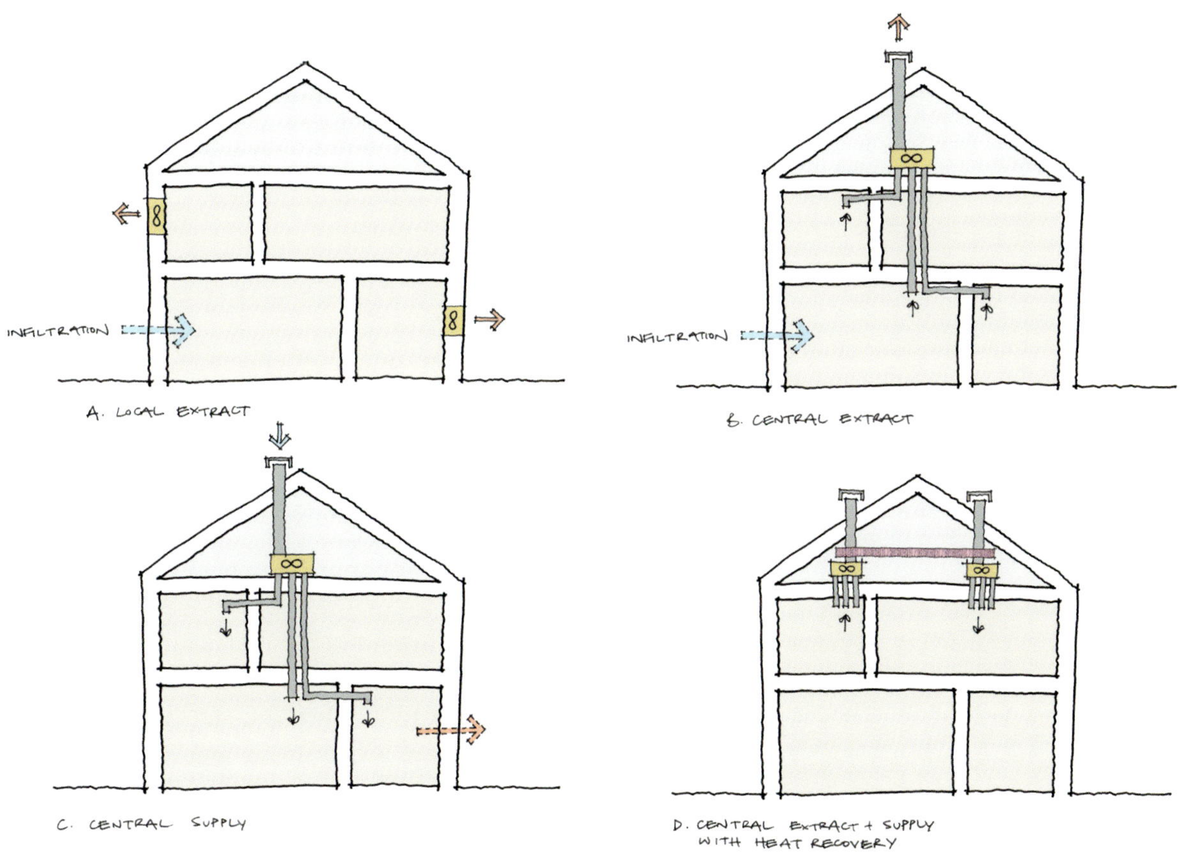

Figure 5.18: Various configurations of mechanical ventilation: (A) local extract, (B) central extract, (C) central supply, (D) extract and supply with heat recovery (MVHR).

in bathrooms, toilets and kitchens. The principle of extracting pollutants – water vapour, odours and cooking fumes – close to their point of production represents sound practice. In the case of cooking, this approach is often enhanced by the use of a ventilation hood over the cooking area in order to reduce the dispersion of the pollutant further (Figure 5.18A). Ventilation rates are high, typically 5–10 ac/h (room volumes) in bathrooms, and 10–15 ac/h in kitchens. These high rates usually mean that the fans are used only intermittently.

In the case of bathrooms and toilets, switch-on is usually initiated by the light switch or an occupancy detector – in a well daylit bathroom, the former will not be effective. Since the bathroom or toilet is intermittently occupied, often for short periods, it is common practice to have a time delay before automatic switch-off, to ensure at least one air change. Humidity detectors are sometimes used; however, these will not detect odours.

For maximum effectiveness, an air inlet must be provided. This could be a gap under the door, in which case the open area should be at least equal to the open area of the fan aperture or duct. Drawing air from the interior will ensure that it is preheated. If air is drawn from outside, it should be at high level to encourage mixing of the cool air in order to avoid cold draughts and stratification.

Kitchen extraction is usually manually controlled, the need being recognised by cooking smells or visible evidence of steam or smoke. Cooking operations such as frying are relatively brief. The considerable noise caused by high rates of ventilation provide an incentive for switch-off. However, some prolonged cooking processes may produce large quantities of water vapour at low rates; a humidity sensor should be used to operate the fan at a low rate and low noise level.

Some extractor fans have electrically or mechanically operated shutters that close when the fan is off. This is to prevent unwanted naturally driven ventilation through the aperture. Passive shutters, relying on wind pressure, are less effective and can cause irritating noise.

Whole-house mechanical ventilation

Whole-house ventilation usually implies the installation of a central fan unit ducted to inlets and/or extract grilles in most rooms. This may be only extract, in which case the extract grilles are located in so-called 'wet areas' – e.g. bathroom and kitchen (Figure 5.18B). This reduces the pressure inside the building, drawing fresh air into the occupied rooms via infiltration and purpose-provided vents if present. The result is much the same as with local extract, whereby the make-up air enters the envelope in the same way.

An alternative configuration is to supply air from a central fan (Figure 5.18C). The room inlets will be located in areas of low pollutant production – e.g. circulation spaces – and low-resistance outlets will be located in areas of high pollution. The building is now at positive pressure. This configuration carries the advantage that the fresh-air (outdoor) intake can be located in a low pollution area and/or the input air can be filtered to reduce gaseous and particulate pollutants. It also reduces the ingress of cold drafts.

The two can be combined to form a balanced system. A further step involves combining the input and extract fans and installing an air-to-air heat exchanger to recover heat from the outgoing air in order to preheat the incoming air, as in Figure 5.18D. This is Mechanical Ventilation with Heat Recovery (MVHR).[32]

MVHR and PassivHaus

MVHR has a number of advantages. It maintains a guaranteed minimum fresh-air replacement rate to ensure long-term air quality. Due to the neutral pressure, the building will be free from casual draughts and infiltration, and due to the heat recovery the input air will enter the occupied space at a temperature close to comfort levels. Finally, the energy cost of the ventilation component of heat loss (see 'Heat loss by ventilation' in chapter 4) will be smaller.

If MVHR is installed in a house with exceptionally high standards of insulation and airtightness, we have in effect reached the PassivHaus standard.[33] This standard seeks to lower both conductive and ventilation heat losses to such an extent that auxiliary heating is zero or very small. The gains from occupants, equipment and solar are sufficient to offset the minimal heat losses, making heating necessary only after prolonged periods of non-occupation or exceptional weather. The smallness of the heating load is often used to justify electrical heating delivered by the ventilation system as warm air, and thus requiring no conventional heating system (Figure 5.19).

Cooling energy is also minimised by the same principles, and through the use of shading and in some cases via pre-cooling of the supply air. Night ventilation and the use of natural cross ventilation through open windows is encouraged during the summer months.

The PassivHaus standard is a comprehensive low-energy standard intended primarily for new buildings. The following energy-performance

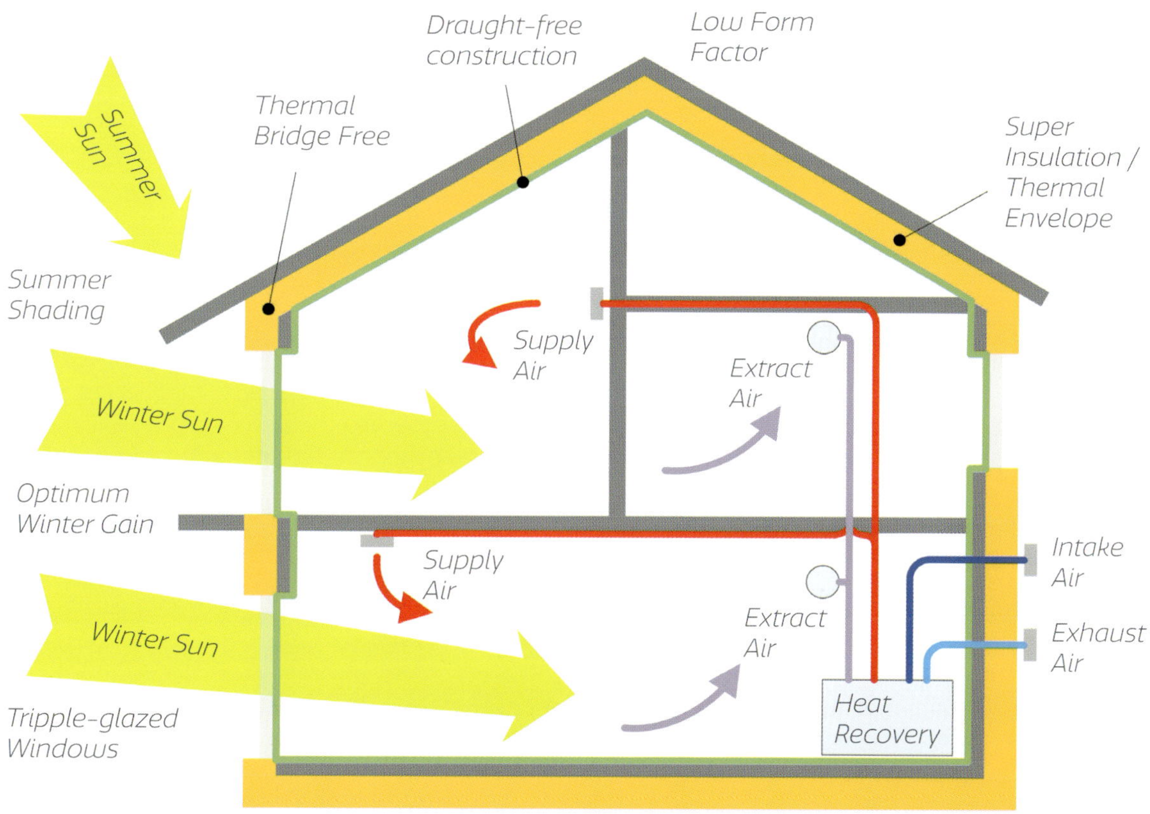

Figure 5.19: The PassivHaus concept showing the two main principles of high levels of insulation and MVHR.

Energy performance targets and air changes per hour	
Specific Heating Demand	*≤ 15 kWh/m² per year*
Specific Cooling Demand	*≤ 15 kWh/m² per year*
Specific Heating Load	*≤ 10 W/m²*
Specific Primary Energy Demand	*≤ 120 kWh/m² per year*
Air Changes Per Hour	*≤ 0.6 ac/h @ 50 Pa (pascals)*

Table 5.2: PassivHaus targets.

targets define the standard, and must be met in order for certification to be achieved.

The PassivHaus standard was developed in Germany in the early 1990s by professors Bo Adamson of Sweden and Wolfgang Feist of Germany, and the first dwellings to be completed to the PassivHaus standard were constructed in Darmstadt in 1991. Since then, this approach has seen considerable interest and application across Europe and in the US. There has, however, been some debate about the cost-effectiveness of the standard, its actual technical performance and the implications for the health of occupants.

Health issues have focused on its high standard of airtightness and the resulting air quality, and the centralised ducted ventilation system. There has also been the perception that the occupant is isolated from the outside, perhaps suggested by the thickness of the insulation. However, there is nothing intrinsic in the standard itself that threatens health. It is likely that misgivings have been prompted by a few poor examples, in which there have been technical problems with the MVHR and controls, and air quality has been compromised.

Airtightness is a measure of the leakiness of a building envelope with 'everything shut'. It need not in any way influence the air-change rate and resulting air quality when the ventilation system is in use – nor when, in milder weather, the operable windows can be opened. But there is a risk that in the event of breakdown or underperformance of the ventilation system, or an unexpectedly high pollution input, the failsafe of infiltration will not be there. Furthermore, the occupier may have developed an attitude that makes them reluctant to use the windows or manual vents, being under the impression that the mechanical system 'should cope'.

There is a view that the ducted air system in the PassivHaus itself creates a hygiene risk. Dust, mould spores and bacteria can develop in damp conditions, which may occur transiently in ducts and filters, and this has often been identified as a cause of Sick Building Syndrome (SBS). This is a potential problem in all ducted-delivery systems, made worse if the air is recirculated – which, in the case of PassivHaus, it is not.

The situation, then, is really no different from any modern building with high insulation and airtightness standards. But this does not diminish the problem; poor air quality – with contaminants such as CO, CO_2, NO_x, VOCs, bacteria, micro particulates, mould spores, dust-mite faecal pellets and other bio allergens – is probably the main agent by which buildings affect our health.

Although heating loads may be very small, it must not be overlooked that the dependence on fans running for a large proportion of the time will create running and maintenance costs, and a possible noise source.

IAQ detection and controls

Mechanical systems offer the advantage of more convenient control than natural ventilation, usually by electrical signals from sensors. Detecting the concentration of a pollutant (or its proxy) and using this to control fan speed or on/off settings is straightforward and should lead to good IAQ.

Whilst humans have evolved a sense of smell that warns of some natural dangers, there are many lethal agents – often generated from new chemicals or technologies – to which we have evolved no sensitivity.

One of these is carbon monoxide, the effects of which have tragic consequences, and it is now commonplace to have CO detectors in place – especially in rooms where solid fuel is being burned. (It is, in fact compulsory under the UK Building Regulations in the case of a new solid-fuel appliance.)[34] However, the unexpected malfunction of a boiler or gas fire has often been the source of fatal CO – and this suggests that detection should be present in all rooms.

Other detectors can be used to control ventilation. Human occupation always creates CO_2 via respiration, which has prescribed limits (1,000 ppm). A control point of, say, 800 ppm will ensure not only that the CO_2 criterion is met but also that other occupant-generated pollutants such as water vapour and odours are also kept within acceptable limits.

The principle of air sampling forms the basis of systems that operate under natural ventilation with mechanical backup. These are referred to as *hybrid systems*. In their simplest, form these might involve a room ventilated by opening windows and manual vents, with an extract fan that operated on a signal from a CO_2 detector. More sophisticated systems might use a passive stack to give whole-house extraction, with a CO_2-controlled damper to prevent over-ventilation and a CO_2-controlled fan in a 'shunt duct'[35] to prevent under-ventilation.

Other control indicators include water vapour, which is also generated by respiration. However, since high humidity can be harmful to the building fabric rather than the occupants, it is more likely to be used as a control point in areas where large quantities of water vapour are produced – such as kitchens and bathrooms.

Control also permits ventilation-demand matching. It is obvious that the actual ventilation rate would ideally respond to occupancy level – a living room with five occupants needs five times more ventilation than if occupied by one, and it could be argued that an empty room needs no ventilation at all. This is rarely the case in practice.

For example, four residents enter an unoccupied living room in which the IAQ is up to a good standard, due to a long period of infiltration. Two hours later, an occupant becomes conscious of stuffiness (high humidity and increasing CO_2 level, combined with body odour or other olfactory pollutant) and opens a window. It is winter, so the motivation is not due to overheating. The open window raises the ventilation rate to well above the steady rate needed. The scenario finishes with the occupants leaving the room for dinner 20 minutes later and failing to close the window – resulting in a waste of energy as the heating system struggles to make up the heat loss, and a chilly room an hour later for the after-dinner drinks! Hybrid ventilation with IAQ control could have avoided this, saving energy and increasing comfort and health.

Key points

- Ventilation has three main purposes:
  1 to maintain minimum indoor air quality (IAQ)
  2 to reject surplus heat
  3 to provide air movement for physiological cooling.
- Natural ventilation is driven by thermal buoyancy (stack effect) and wind pressure. Both are highly variable, which gives rise to widely fluctuating ventilation rates unless actively controlled.
- IAQ is affected by pollutants generated in a building by its occupants, contents, activities and the quality of the outdoor air.
- Indoor finishes and materials should be chosen to have minimal pollutant potential.
- The effects of pollutants can be mitigated by
  - reduction
  - dilution
  - extraction.
- Extraction of pollutants should occur as close to the point of production as possible.
- Well-sealed cupboards ventilated to the outside
  - can reduce indoor pollution from clothes airing and storage.
  - should be provided for storing dangerous household substances such as bleach, fabric-cleaning fluids and oven cleaner.
- Air intake for both natural and mechanical systems should be located away from pollution sources, such as road traffic.
- Mould growth and bacteria
  - occur at high humidity even though surface condensation may be absent
  - are associated with serious health effects such as asthma and other allergies.
- Mould growth is unsightly, smelly and damages materials – and is not conducive to wellbeing.
- Mechanical handling of intake air offers the opportunity for filtration of particulates and some gaseous pollutants.
- Mechanical handling of intake and rejected air offers the opportunity for heat recovery (MVHR).
- The PassivHaus Standard combines MVHR with very high levels of airtightness and insulation, and a target of very low energy consumption.
- The use of pollutant concentration – e.g. CO_2 – as a control point for natural or mechanical ventilation can improve IAQ and prevent energy waste, and is an essential part of hybrid systems.

The Case for Daylight

Sunlight and daylight have a long history of being associated with health, as shown in chapter 1. Hippocrates, the father of modern medicine, recommended sunbathing. A solarium, a room open to the south, was built into the ancient Asclepieon (the forerunner of the hospital) at Kos in around 350 BC. More than two millennia later, sunlight was again espoused as a health-giver. Exposure to it was recommended for the treatment of TB; the term *heliotherapy* was coined. Lack of sunlight in poor urban living conditions was (correctly) blamed for the bone disease rickets.

In a book, *School Architecture* written in 1874, E.R. Robson, consultant architect to the UK Education Department, wrote, 'It is well known that the rays of the sun ... are to a young child very much what they are to a flower.'[36] Though lacking in technical rigour, this poetic statement seemed to have had much influence in the design of windows in schools for the next century. Ironically, this trend culminated with the over-glazed schools of the 1960s that, due to their environmental shortcomings, marked the beginning of a decline in enthusiasm for sunlight.

More recently, overexposure to sunlight has been associated with skin cancer. The response to this has been alarm - good for the manufacturers of skin creams, sunshades and sunglasses - and the provision of sun protection is now being enshrined in building regulations for schools.

However, we are currently in a situation where the benefits and dis-benefits of sunlight and daylight are both recognised, and the resolution of compromise is being based on scientific evidence rather than dogma. This is typified by the INTERSUN Programme of the WHO, which was prompted in 1992 largely as a response to the damage to the Earth's ozone layer and the realisation that UV levels were on the increase.[37]

Health and wellbeing issues have already been outlined in the first two chapters. Here, we are mainly concerned with the design response to the need for good daylighting and sunlight penetration and, at the same time, providing control to prevent their negative aspects.

The sky as a light source for the room

Daylight and sunlight come ultimately from the same source - the sun. Figure 5.20 shows how the intensity and spectral content of sunlight is influenced by its transmission through glass. We usually refer to *daylight* as being light from the diffuse sky (cloudy or clear) and *sunlight* being direct from the sun's disc. The light that falls on the earth's surface covers a wide range of frequencies (or wavelengths).

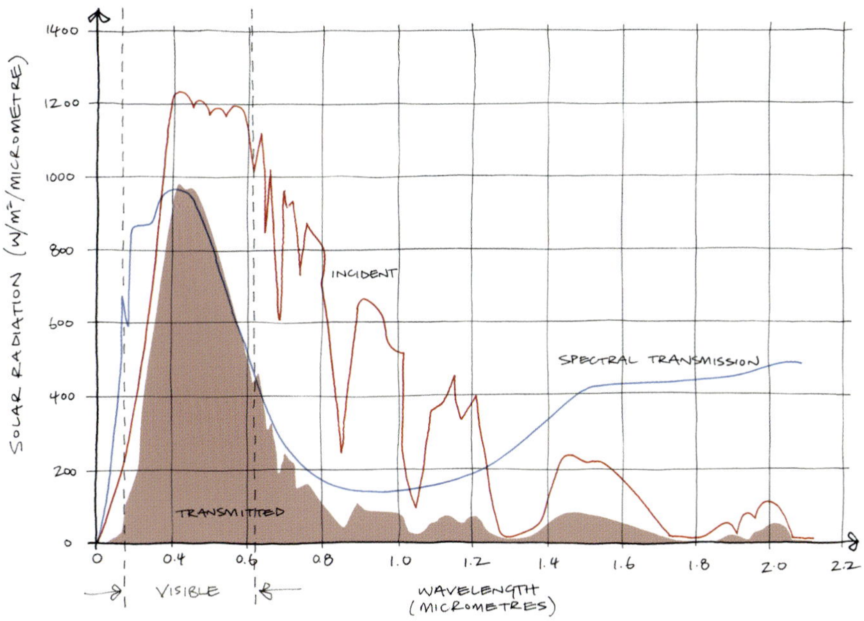

Figure 5.20: The spectrum of incident solar radiation and radiation transmitted through glazing, showing the visible region.

Our eyes have evolved the greatest sensitivity to the most intense part of the spectrum. However, there is an equal amount of energy in total outside that visible spectrum, in the ultraviolet (UV) and the infrared (IR) regions.

Ultraviolet light

Known physiological effects are associated with the UV and the 'blue' end of the spectrum – both of these are present in direct sunlight. Scattered light from the diffuse sky generally contains a greater proportion of these shorter wavelengths, but at lower intensity than that of direct sunlight. More specifically, UV is divided into UV(A) (longer wavelength), and UV(B) (shorter wavelength), and the known physiological effects are shown in Table 5.4 (see page 149). The more damaging UV(B) is absorbed by all types of window glass. Some UV(A) is transmitted by clear and blue glass, but totally absorbed by green- and brown-tinted glass and by solar-control films.

Visible light

The visible spectrum carries information – the view out of the window towards the countryside or the lines of text in a newspaper. Inadequate light indoors for visual tasks such as reading can result in eyestrain and temporary discomfort, but there is no evidence that it causes permanent eye damage. On the other hand, inadequate daylight has been shown to have negative psychological effects.[38]

However, the cheapness and convenience of electric light (together with two major leaps forward with fluorescent, and now LED lighting) means that daylight is all too often abandoned as the working illuminant, and even in small rooms such as domestic kitchens some form of artificial light will be used even in daytime. Whilst the artificial illumination may satisfy technical-illuminance criteria, three factors result in it being inferior to daylight:

1. Few artificial-light sources have a full continuous spectrum, rich in blue, essential for circadian entrainment (see 'Daylight and the circadian rhythm'.
2. The dynamic nature of daylight in both intensity and colour reinforces the circadian cycle.
3. Apart from providing daylight illumination, windows also usually afford an external view – often including glimpses of nature.

A further aspect of indoor daylight is its role of revealing and enhancing occupied spaces. This is difficult to quantify, but architects are very familiar with this concept. However, everyone is at some time delighted by the way in which bright sunlight reveals the geometry and texture of the surfaces of a room or the soft light of a north-facing rooflight bathes a space in gentle graduations of luminance. This function performed by electric lighting is usually static and may seem contrived and artificial.

Infrared

Near and far infrared (IR) have no direct visual or physiological effects other than warming the skin and clothing, and warming a room. Often known as heat radiation, it forms the largest component of solar radiation that causes heat gain. However, it must be remembered that the visible part of the spectrum also creates heat gain.

Light transmission through windows and the Daylight Factor

How do we design to get adequate sunlight and daylight into our rooms? And what is sufficient, and what is too much?

The brightness of the sky and the resulting illumination falling on the ground or a building is highly variable. Thus, in describing how much ambient light enters a room it is convenient to use the ratio of illuminance inside to the unobstructed illuminance outside the room. This is the Daylight

Factor (DF), as shown in Figure 5.21.

The DF has three components, as shown in Figure 5.22:

1. the Sky Component (SC)
2. the Externally Reflected Component (ERC)
3. the Internally Reflected Component (IRC).

All of these are in turn dependent on three factors, all of which are in the control of the designer (to a greater or lesser extent):

1. the geometry of the building
2. the transmission of the glazing
3. the reflectance of the surfaces.

In the case of the SC, the key parameters are the position, shape and size of the window(s) together with the presence of outdoor obstructions affecting the view of the sky. The ERC is influenced by the geometry of any obstruction and, crucially, its reflectance. A further factor is the illumination of the obstruction.

Both SC and ERC are directly influenced by the transmittance of the window. This has already been referred to in 'Sunlight and daylight access' in chapter 4, in relation to orientation. Transmittance will also be influenced by treatments to the glass such as tinting, reflective and low-e coatings. Lastly, further to earlier comments concerning obstruction due to glazing bars, there is a growing trend for windows to have more of these – often in order to reproduce the look of a bygone era. Furthermore, double glazing and thermal-break design has demanded thicker frame sections. This can lead to 50% or more of the building aperture being obstructed, by framing and bars, and may well go unnoticed in the design process.

The ERC can be very significant. Consider a north-facing window that overlooks a south-facing white-painted wall. When the sun is of sufficient elevation to illuminate the wall, its brightness will be considerably greater than that of the unobstructed blue or clouded sky providing the major component of daylight. If, however, the wall is a dark colour, the main source of daylight for the room would be greatly reduced. This phenomenon is relevant to high-density housing – particularly courtyard layouts.

Equally important is the IRC. This is dependent on the average reflectance of room surfaces; the external ground reflectance; and, to some extent, the spatial distribution of reflectances within the room. In

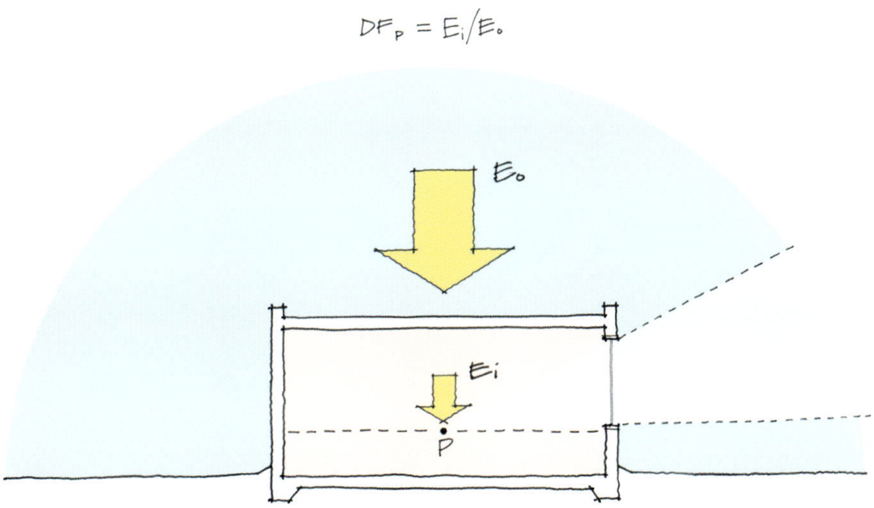

Figure 5.21: The definition of Daylight Factor: DF = ratio of internal illuminance (E_i) at a point to illuminance from the unobstructed sky (E_o) – thus, DF = E_i/E_o (%).

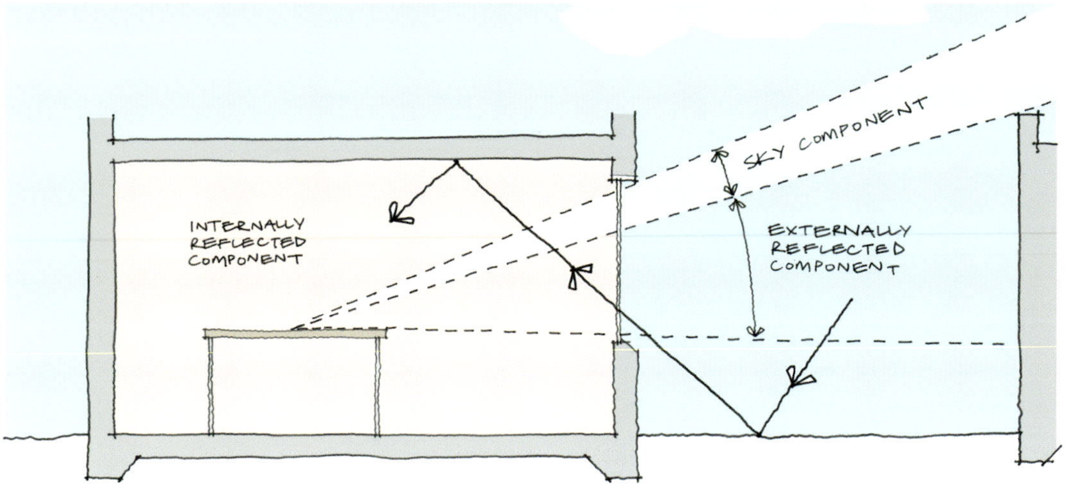

Figure 5.22: The three components of the Daylight Factor.

specifying finishes, it is good to consider the path via multiple reflections that light will take. It is evident from Figure 5.22 that the reflectance of the ground and ceiling is crucial. Furthermore, because the side walls get a view of the sky the reflectance of the side walls also matters.

It is useful to remember that, at any point in the room, the SC only exists where there is a view of unobstructed sky. The ERC only exists only where there is a view of the window. But the IRC exists everywhere. This should be borne in mind when planning layouts for irregular-shaped rooms. Another issue is that because a person's normal view is horizontal, the walls form the largest area in the field of view. It follows that the reflectance of the walls plays an important part in the perception of the brightness of a room.

Finally, a vital design element is the provision of shading devices, which range from simple curtains (primarily to provide privacy) to sophisticated devices that allow modulation and redistribution of daylight. This is discussed in 'Sunlight penetration and shading' in chapter 5.

The DF has an interesting history. It was developed in the pre-computer era as a parameter that could, with some effort, be calculated from design information (geometry, reflectances, etc.). To simplify the procedure, tables and graphical methods were devised. With the coming of computers in the 1980s, these methods were replaced by software, enabling multiple calculations to give the spatial distribution of DF as shown in Figure 5.23. More recently – with the growth of computing power and memory, and better characterisation of sky luminance – it is common for daylight-modelling software to output daylight calculations with actual illuminance values (in lux), rather than DF. In the authors' view, this is a retrograde step as it combines two unrelated parameters, DF and sky luminance, with only the former being a design variable. DF remains a good conceptual parameter – a measure of a building's transparency to the sky. It is of particular value in an educational situation, where concepts and principles are being developed.

Standards, on the other hand, mainly serve practice – and it is here that the end result is on trial. It is then more appropriate to frame the standard in terms of achieved luminances, integrating the climate with the building design. This is the approach taken by the Daylight of Buildings Standard EN 17037[39] – combining lux levels with median DF so as to reflect the climate-based dependence of daylight on latitude. It will also set standards for view, sunlight penetration and glare.

The notion of *daylighting autonomy* is the degree to which daylight provides sufficient illumination without artificial light, usually expressed as a fraction of time – e.g. occupied hours. This is a useful concept in the working environment, where the task and occupancy are well defined. The concept could be adapted using illuminance and time criteria set by circadian-synchrony requirements (as discussed in 'Daylight and the circadian rhythm' later in this chapter) in order to make it more applicable to the home.

Actual lighting levels in rooms

From the definition of DF, the internal horizontal illumination by daylight E_i at any point in the room can be calculated by:

$$E_i = E_o \times DF$$

where E_o is the outside horizontal illuminance.

We have already established that the illumination from the sky is very variable and is best described by cumulative frequency curves as in Figure 5.24. These show the fraction of a certain period (usually the working day or sunrise to sunset) that the illumination from the sky is above a certain level. The DF can be used to transform this to the fraction of the same period that the interior illumination (at that point in the room) is above a value.

The example shows that the position A in the room where the DF is 2% will achieve 300 lux for

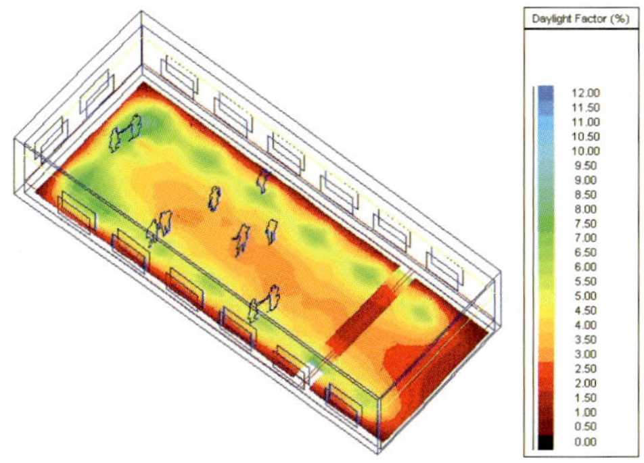

Figure 5.23: Spatial distribution of DF in a room generated by daylight-design application FlucsDL (IESVE software).

63% of daylight hours, but 1,500 lux for only 8% of daylight hours.

However, the DF is also very variable spatially. The position nearer to the window, B, where the DF is 15%, will receive 300 lux for 90% and 1,500 lux for 70% of the time. The fraction of time that a zone in a room spends above a certain illuminance value is relevant to the exposure to daylight of an occupant for circadian synchrony (see 'Daylight and the circadian rhythm' later in this chapter). This is relevant in the case of residential buildings where occupancy density is quite low and occupants are usually free to move about. It may not be the case in homes for the elderly and/or occupants of limited mobility.

Sunlight penetration and shading

Direct sunlight carries about five times more energy than diffuse light from the blue sky, thus its impact on the occupant is much greater. This can include its heating effect,[40] visual discomfort or injury by the UV content. Direct sunlight obeys easily predictable geometry, so that in principle the exact position of sun patches on wall and floor surfaces can be predicted for any time of day and season. Most environmental-design software tools include sunlight-penetration visualisation,

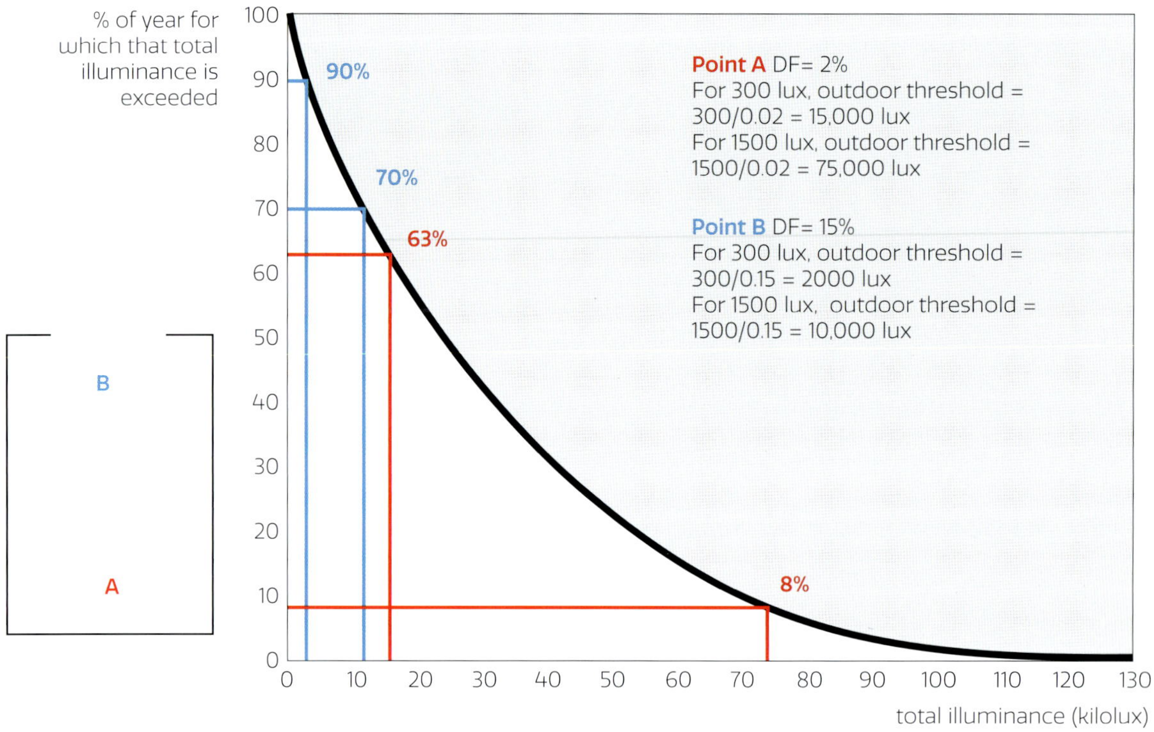

Figure 5.24: Cumulative frequency curve for percentage of time sky illuminance is above a datum value during daylight.

and some will predict hours of exposure from climate-derived solar data.

Whilst direct sunlight has benefits, such as contributing to space heating as described in 'Internal heat gains and energy balance' in chapter 4, it also has the dis-benefit of causing overheating in the warm season. The health implications of overheating can be serious, and have already been discussed in 'Extreme temperatures and health' earlier in this chapter. Solar radiation has many benefits to health – in particular, the production of vitamin D and the effect of synchronising the human circadian rhythm (see 'Daylight and the circadian rhythm' earlier in this chapter) – but even after passing through glass, it still contains near-UV(A), which can for some very delicate skin types cause sunburn and even skin cancer. This dual aspect of sunlight means that shading is essential.

Common shading devices include the following:

- fixed overhangs
- louvres – fixed and adjustable, vertical and horizontal
- curtains
- roller blinds
- fixed perforated screens and grids
- tinted and reflective glass
- photo-chromic and switchable glass.

Louvres and overhangs rely on geometry to prevent transmission from the part of the sky through which the sun passes, whereas the remainder rely on reduced transmissivity.

Shading type		Orientation	View	Natural ventilation (in limiting conditions)	Daylight (in limiting conditions)	Seasonal response	Modulation	Notes
Overhangs 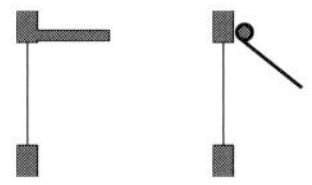	*fixed*	*180 +/- 30*	*good*	*good*	*medium*	*medium*	*none*	
	retractable	*180 +/- 30*	*good*	*good*	*good*	*good*	*good*	*e.g. canvas awnings + adjustable geometry*
Louvres 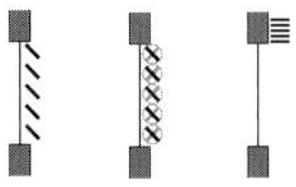	*fixed*	*180 +/- 30*	*medium-poor*	*good*	*medium*	*medium*	*none*	*view influenced by blade module size & geometry*
	adjustable	*all*	*medium-poor*	*good*	*good*	*good*	*good*	
	retractable	*180 +/- 30*	*medium / good*	*good*	*good*	*good*	*medium*	
	retractable + adjustable	*all*	*medium /good*	*good*	*good*	*good*	*good*	*'good' applies to when retracted*
Fins (vertical) 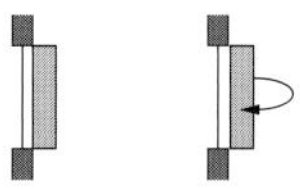	*fixed*	*90, 270 +/- 20*	*medium-poor*	*good*	*medium*	*medium*	*none*	*view influenced by blade module size & geometry*
	adjustable	*90, 270 +/- 45*	*medium*	*good*	*good*	*good*	*good*	
Blinds 	*retractable*	*all*	*poor/good*	*poor*	*good*	*good*	*medium*	*'good' applies to when retracted*
Perforated screens 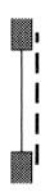	*fixed*	*all*	*poor*	*medium-poor*	*poor*	*poor*	*none*	*not recommended*

Table 5.3: Shading device types and performance.

A summary of shading types is given in Table 5.3. For detailed design advice on shading, refer to specialist texts. Here, we outline a few issues that may impact on health and wellbeing. Two important issues are the impact of shading devices on glare, and view.

> Notes for using table 5.3:
>
> Orientation *all* implies that performance is not orientation-sensitive – although in general, there would be no demand for shading on façades orientated towards the poles (north or south) +/- 45°. Adjustable louvres and fins will have poor *view* performance and poor *natural ventilation* performance, when completely closed.
> *Natural ventilation (in limiting conditions)* refers to situations requiring the shading to be deployed – i.e. when there is an overheating risk. *Good* implies that natural ventilation is not impeded.
> *Daylight (in limiting conditions)* refers to minimal daylight availability when there would be no requirement for shading. *Good* implies that the shading devices cause no reduction in daylight transmission in these conditions.
> *Distribution* refers to the ability of the shading device to improve daylight distribution, thereby lowering the total daylight energy (solar radiation) required.
> *Seasonal response* refers to the ability of the shading device to respond to different sun angles at different seasons.

Glare

Glare is due to three factors:

1. the luminance (brightness) of the glare source relative to the background
2. the size of the glare source – as a proportion of the total field of view
3. the probability that it will fall into the subject's field of view.

Glare is divided into two categories, *discomfort glare* and *disability glare*, terms that are self-explanatory. Both clearly have negative effects on the health and wellbeing of occupants.

The most common source of glare to indoor occupants is a sun patch, whereby direct sunlight falls on a wall, floor or furniture surface that is in the field of view. The larger the glazed aperture, the larger the patch; the greater the reflectance of the surface, the brighter the patch will be. This can be a problem in functional spaces such as an office or studio, where seating position and orientation may be constrained by the nature of the task and by occupancy density.

Specular reflection from shiny surfaces can also cause glare because, due to the mirror-like reflection geometry, there is a high proportion of luminous energy in one direction. Thus, even though the total reflectance of, for example, a polished wooden table may be only 20%, if half of that is specular, (due to the polish), 10% of the luminance of a direct sunbeam is still very high – and will be concentrated in one direction.

The shading device can itself become a glare source. Light-coloured translucent blinds can become highly luminous when illuminated by direct sun. Also, if horizontal or vertical louvres of high reflectance are adjusted in such a way that the sun-illuminated surface is visible, they too can become glare sources.

However, in a residential situation where there is more opportunity for the occupant to change position or orientation, and if adjustable shading devices are available and their use understood, it is less likely that prolonged discomfort from glare will be experienced.

Interference with view

Different shading devices have varying impacts on view, as indicated in Table 5.3 – from almost no impact in the case of an overhang or tinted glazing, to a maximum (i.e. no view) in the case of the opaque roller blind. Somewhere in the middle of this range lie horizontal or vertical louvres, which allow some view.

It is better to have deployable devices – i.e. those that can be placed in or removed from the aperture, completely – even if when deployed they cause some interference. Fixed devices give the occupant no choice – they have to look out at the view through them, irrespective of whether they are needed or not. Fixed devices such as grids and perforated screens give poor shading performance (no geometric selectivity of directional sunlight – unlike louvres), seriously degrade the view and generate a feeling of isolation from the outside. Some people may even find the periodicity of the structure migraine-inducing. Fixed grids and screens should be avoided at all costs.

Glass, printed with varying sized dots in various colours, sometimes referred to as *fritted*, has enjoyed some popularity amongst architects. However, it degrades the view and cannot be adjusted, and should only be used where there is permanent need for a visual privacy function.

Daylight and the circadian rhythm

We have developed a physiological rhythm that controls our alertness and sleepiness to coincide with a normal 24-hour cycle of activity. This is initially controlled by a hormone, melatonin, which is produced by the pineal gland in the brain as a result of certain stimuli. Increased levels of melatonin create drowsiness and aid restorative, healthy sleep; wakefulness is associated with low levels of melatonin. This is referred to as the *circadian rhythm*, from the Latin *circa* ('around') and *dies* ('day').

The main environmental stimulus for melatonin suppression – i.e. to 'switch off' drowsiness – is light. It is generally accepted that exposures to light of around 1,000 lux are necessary, but there is some evidence that values as low as 200 lux may be sufficient for prolonged exposure. As well as intensity, the spectral content of light is also important. It has been found that white light rich in blue is more effective at suppressing melatonin production – i.e. prompting wakefulness – and light rich in red at lower intensities allows melatonin production in preparation for sleep.

This sensitivity to light is an evolved response from when our lives were spent largely outdoors, and when artificial light was rare, of low intensity and had little impact. It is generally held that it had survival value in preparing primitive humans for hunting and gathering activities, as well as defence. The low and red-rich illumination of sunset and firelight became the signal for a reduction of activity in preparation for sleep.

If a person is isolated from a regular stimulus, their cycle of melatonin production still occurs but drifts to a natural period of about 25 hours. This has been experimentally proven by isolation experiments on human subjects. When this occurs to an individual, they will quite soon experience low levels of alertness in the day and poor sleep due to wakefulness at night, and it will have long-term effects on their health and wellbeing. It is the same cause as the familiar effect of jetlag, in which changes of local time are the cause of the mismatch.

The phenomenon of synchronisation by a stimulus such as light is called *circadian entrainment*. Figure 5.25 shows that its effectiveness is sensitive to light intensity and the time at which it is experienced. The simple outcome is that to achieve good entrainment, people should experience large doses of blue-rich light at the start of the working day. The most

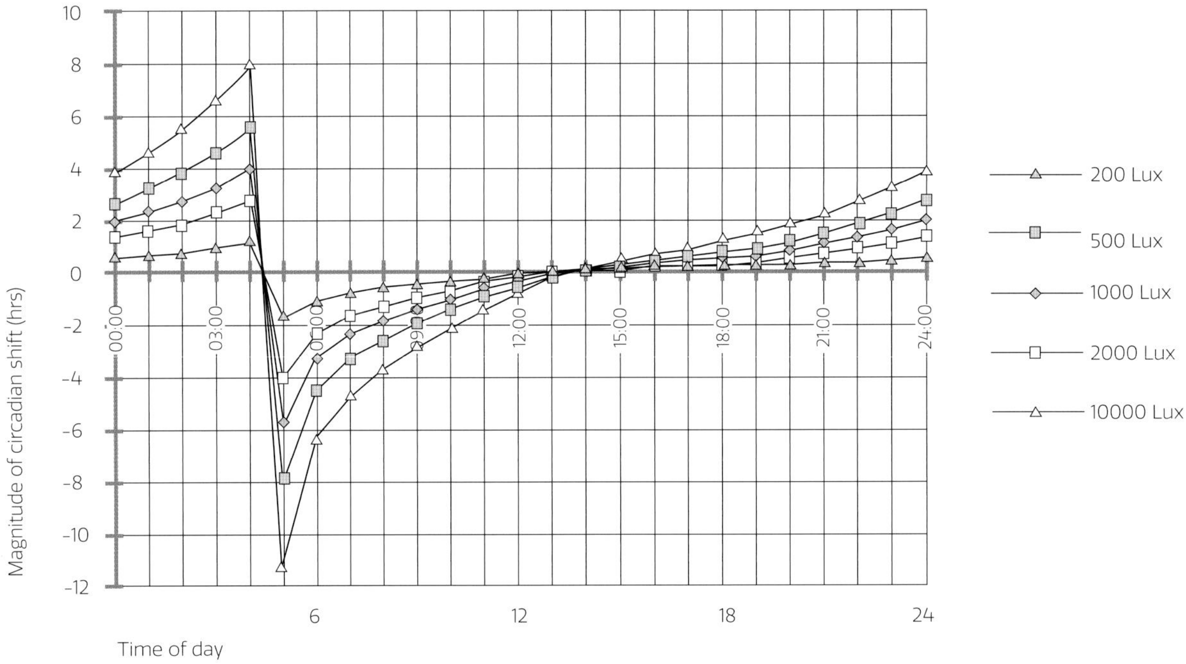

Figure 5.25: Effectiveness of circadian entrainment as a function of intensity and time of day. The discontinuity at 4am is where bright light exposure moves from delaying sleep to arousing the individual for the coming day.[41]

reliable source of high intensities and spectral content is, of course, daylight. The converse of this is that high levels of blue-rich light late in the day are poor preparation for restful sleep.

There are, however, many uncertainties. For example, what does 'an illuminance of 1,000 lux' really mean? Is it on the workplane (the normal specification)? Is it on the occupant's face? Or is it related to the illumination of the retina, where the non-visual receptor is located. If the last-named is true, then it is the existence of a view of bright surfaces (such as the sky) that matters. However, as wide-angle views of the sky will also result in high illuminance levels the two parameters are interrelated.

There is a wealth of evidence that deprivation from daylight has negative effects on health and wellbeing, summarised in the BRE interim report describing an initial literature review on circadian lighting and workshop to review key issues.[42] Impacts range from well-established evidence of sleep disruption[43] and poor daytime mood, concentration and alertness,[44] to correlations between incidence of cancer in night-shift workers,[45] as well as obesity There is also evidence that daylight may influence recovery from illness and surgery,[46] but is this a circadian effect? It is often complicated by the combination with view, usually a natural view. This issue has already been discussed in 'The case for access to nature' in chapter 3.

We have attempted to summarise the health effects of light, together with the key technical parameters, in Table 5.4.

Note that several symptoms can be attributed to circadian asynchrony but could also be caused by associated factors.

In a study carried out by Douglas Cawthorne,[56] the vertical illumination on the face of subjects was monitored using a sensor mounted on a band worn on their foreheads, recording onto a portable data logger. Two groups of subjects were chosen, one occupying a shallow-plan 'well daylit building' and one in a deep-plan, predominantly artificially lit office building. The results demonstrated a dramatic difference in light exposure of the two subjects, with facial illuminance peaks of 3,000–8,000 lux compared to consistently below 200. It is interesting to note that the modern deep-plan office would not have been described as 'poorly lit'. It is simply that in the well daylit building, the many vertical windows provided very high facial illuminance locally, and bright sky views.

To put some figures to that: a typical artificial illumination on a vertical surface would be 200 lux. But standing close to a window, where the vertical DF could easily be as high as 20%, facial illuminance even on an overcast winter's day would be 1,000 lux – and could be as high as 10,000 lux for a sunny day.

Health effect	Active Agent	Mechanism	Reference
Vitamin D production	*UV(B)*	Outdoor *sunlight – unspecified number of hours between 11.00 and 15.00 hrs exposed arms and legs without sunscreen between April and October*	*NHS Choices website*[47]
Skin cancer	*UV (A & B)*	Outdoor *sunlight* Indoor *sunlight through clear glass – unspecified time, dependent on skin type and intensity, sunburning indicates risk*	*Cancer Research UK website*[48]
Germicidal effect	*UV(B)*	*Outdoor or glassless exposure to direct sunlight – damages DNA*	*Weinziri, 1914*[49]
Prostate cancer and breast cancer	*Light at night*	*Night working, lack of daylight causing lack of circadian synchronisation*	*Kloog, et al., 2009*[50]
Poor daytime alertness	*Lack of morning exposure to blue-rich light*	*Lack of circadian synchronisation, melatonin production extended*	*Bedrosin and Nelson, 2013*[51]
Disrupted sleep	*Daylight, sunlight, view*	*Melatonin production prevented*	*Sleep Council 2013*[52]
Quicker recovery from surgery	*Daylight, sunlight, view*	*Circadian synchronisation + possible psychological effects*	*Ulrich, 1984*[53]
Depression	*Lack of indoor daylight and sunlight*	*Not known – possibly poor circadian synchronisation and/or vitamin D deficiency*	*Kent, et al., 2009*[54] *Penckofer, et al., 2010*[55]

Table 5.4: Health effects of sunlight and daylight.

Cawthorne's research was carried out in the workplace – specifically, a university chemistry department (the daylit space) built in the 1930s, and a 1990s-built deep-plan office. In the office building, the occupants were much more sedentary and this, in combination with a large proportion of the floor plate receiving very little daylight, would account for the very low light exposure.

Design response

Translating this to a domestic situation, it is reassuring to note that all residential buildings are relatively shallow plan, and that regulations and codes of practice prescribe minimum DFs requiring glazing areas that will result in areas of high illuminance close to windows (although regulations also recommend maximum glazing areas – 15% of the floor area – to limit heat loss). Assuming occupants of normal mobility, there is a high probability that they will occupy zones of high daylight illuminance from time to time. There is of course no guarantee that they will visit these zones, which suggests that room layout and furnishing could have an impact.

An obvious example is that combining a window seat with a pleasant view would greatly increase the probability of high levels of daylight exposure. Conversely, a deep narrow living room with circulation space at the daylit end might result in the occupants retreating into the gloom of the television and the soft furnishings.

We must bear in mind that it is the retinal illumination that ultimately affects circadian entrainment. This means that the brightness of surfaces in the field of view will have a significant impact. Surface brightness is dependent on the product of illumination and reflectance. Thus, the use of light finishes for walls will have double significance – not only increasing the internally reflected component (IRC), adding to the overall DF, but also providing a visual field with high luminance.

As hinted at above, we also have to consider occupant mobility. Frail and elderly people may spend many hours sitting in a favourite chair, and if this is in a poorly daylit area it may commit them to weeks of underexposure. On the other hand, direct sunlight can create thermal discomfort and glare. In the presence of adaptive opportunity – i.e. a flexible room layout with controls such as curtains or blinds – these conflicts can often be managed by fellow occupants or carers. But this is usually out of the architect's control.

Bedrooms create their own set of problems. The circadian-entrainment principle means that it is desirable to have bright, blue-rich daylight early in the morning to prepare for wakefulness. But how early? In middle latitudes, sunrise at 09.00 hrs in the winter may be too late and 04.30 hrs in the summer is definitely too early! What is our design response to this? Some individuals may be more sensitive to light in terms of sleep disturbance than others, and may require total blackout. This can be achieved by blinds or curtains of suitable specification, and is commonly found in hotel rooms in high latitudes. Total blackout does not mean that the blinds or curtains need to be of dark material, as shown in Figure 5.26.

Having access to natural light upon waking, when it is available, suggests that the bathroom, circulation space and kitchen of a home would benefit from being daylit so as to offer that early-morning circadian opportunity (Figure 5.27). Such spaces tend to be overlooked when it comes to daylighting design, or relegated to the middle of house plans. The bathroom is probably the first space a person would use upon awakening – to wash, brush teeth or apply makeup. Some have even made reference to taking a 'shower of light' in a well-lit bathroom. This is followed by moving via the corridor or stairs and then the kitchen, to prepare breakfast or grab a morning coffee. Each provides a

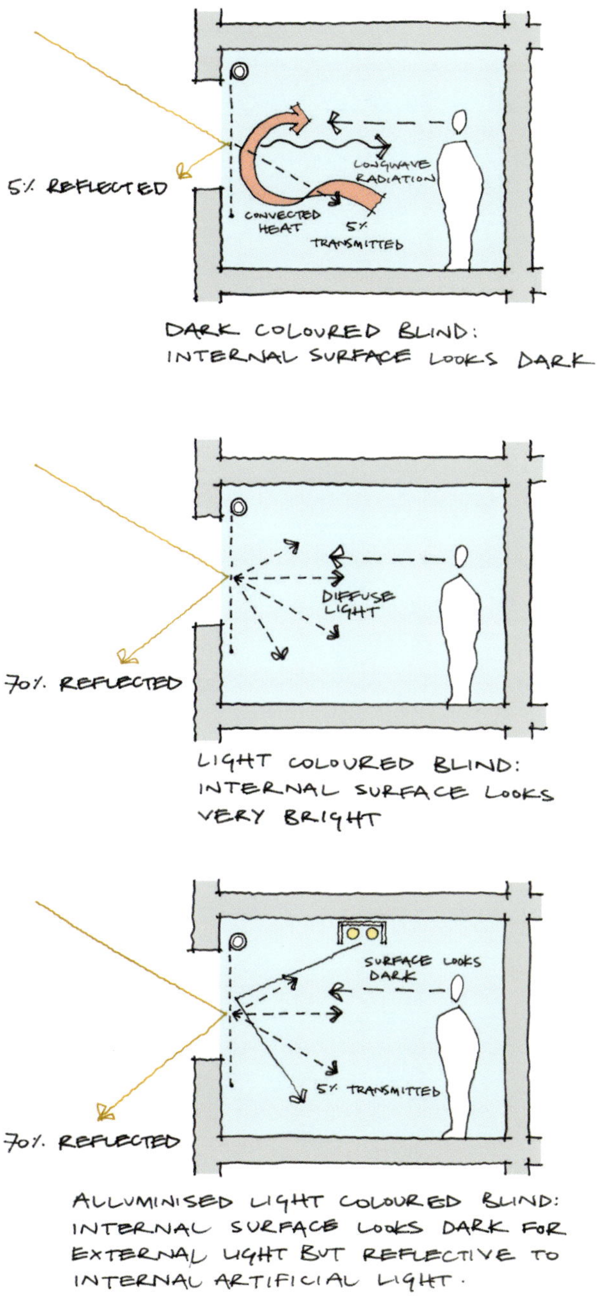

Figure 5.26: Photometric function of various blind materials.

potential opportunity for daylight exposure, particularly for those who are constrained to or work from home.

It is often held that bedrooms should be orientated to the east and living rooms to the west. But without blackout, this could exacerbate the problem of early summer sunrise, and without shading could lead to overexposure to bright sunlight late in the day as well as overheating. We must also bear in mind that precise orientation is only significant on sunny mornings (say, between the hours of 06.00 and 09.00). The diffuse sky is predominant in mid-Europe – particularly in winter, when sunny mornings typically occur only one day in six.

Furthermore, this prescriptive orientation is a major constraint in planning terms, and may well conflict with other environmental factors such as noise or view. Bearing in mind these potential conflicts and a lack of certainty over the extent to which entraining light exposure can be predetermined by design as distinct from behaviour, the authors feeling is that a rigid-circadian orientation principle should not be pursued.

Rather, the designer should provide a flexible design, rich in adaptive opportunity, responding to the unique potential of the site. Hopefully, an understanding by the occupant of how to use it to promote health and wellbeing will become part of their own education, similar to the pursuit of a healthy diet and sufficient exercise. The designer could contribute to this in the explanation of their design, drawing attention to these issues and including them in the handover documentation.

Figure 5.27 overleaf: Morning routines from the bathroom, via the stairs to the kitchen – each provide opportunities for a healthy dose of daylight.

Key points

- Sunlight and daylight come from a common source, and have a similar (but not identical) continuous spectrum. Daylight is diffuse, from all parts of the sky, and sunlight is direct and can be predicted by straight-line geometry from the sun's disc.
- Daylight intensity in a room is dependent on the transmittance and area of unobstructed glass, not just the area of a built aperture in the envelope.
- There is long-established evidence that sunlight within buildings is beneficial to human beings, but it also has dis-benefits.
- This suggests that shading controls, allowing the occupant to modulate the penetration of sunlight and daylight, should always be provided for windows receiving direct sun.
- Some selective control of direct sunlight can be achieved by the orientation of fenestration and fixed obstructions, overhangs or fins.
- One of the functions of daylight and sunlight is to synchronise the circadian rhythm. This requires morning exposure to bright daylight. However, it is not recommended that orientation of the building to morning sunlight is used as an overriding principle.
- Shallow-plan rooms with adequate glazing area, together with modulating shading controls, can provide high and low illuminance to cater for circadian synchronisation and healthy sleep.
- Special consideration should be given to light distribution and control for old and infirm occupants with limited mobility.

Artificial lighting

Over the last two centuries, the main form of domestic lighting has moved from candles; to oil; to gas and incandescent mantles; to electric filament (tungsten); to fluorescent; and, finally, to LED. In that development, luminous efficacy has increased from about 0.5 to 150 lumens per watt – a 300-fold increase. At the same time, energy costs, allowing for inflation, have reduced drastically; even if we apply a modest factor of 10, we can see that a lumen of artificial light now costs 1/3,000th what it did 200 years ago. Furthermore, unlike the naked flame of the 19th century it presents no fire or local pollution hazard, and its colour can be closely controlled.

This has had both beneficial and negative effects. In the workplace – following the leap forward in luminous efficacy of the fluorescent lamp in the 1950s, and to a large extent continuing today – daylight has been all but abandoned as a working illuminant. This allowed deeper plans, with workstations (other than those at the perimeter) relying totally on artificial light. Typical illuminances of 300–400 lux, although adequate for the visual task, were below values that become fully active in circadian synchrony. Furthermore, poor colour rendering and 50 Hz flicker, together with loss of outdoor views, resulted in this lighting environment being far from healthy.

Fortunately, this dependence on artificial light has never reached housing to the same extent. This is largely due to the shallower plan, making daylight available to nearly all habitable spaces. Furthermore, this issue is covered by regulations in terms of the window/floor area ratio.

However, artificial light is often used to supplement daylight during the daytime and, due to the evening being the most active occupation period, occupants spend a large amount of time in artificial light. This section looks at possible health implications of residential lighting in terms of spectral content and colour rendering, luminaire design and light distribution, and controls.

Spectral content and colour rendering

To be active in suppressing melatonin production – i.e. raising alertness – light must be blue-rich. Although this is often referred to as 'cool light', it is more accurately described as 'high colour temperature light', referring to the temperature of a black body that would emit light of a similar visual sensation. Typical values, suitable for daytime illumination, would be around 6,000 K. Light active in stimulating melatonin production, and hence lowering alertness, is at the other end of the spectrum – i.e. rich in red. Often confusingly referred to as 'warm-white', this is technically described as 'light of low colour temperature' – typically, around 2,700 K.

Both compact fluorescents and LED lamps are available in three broad categories:

1 soft white/warm white (2,700 K–3,000 K)
2 bright white/cool white (3,500 K–4,100 K)
3 daylight (5,000 K–6,500 K).

Colour rendering is a measure of how consistently and faithfully an illuminant reveals the colour of a surface. Colour temperature is not a determinant of colour rendering; the latter is influenced by the continuity of the spectrum. Whilst the visual sensation of any colour can be synthesised by three components of monochromatic (primary) colours, this is not so in the case of viewing pigments.[57] If the pigments have specific spectral reflectances and absorptive wavelengths (which most do), viewed by reflected light their apparent colour may be distorted if the wavelengths of the illuminant do not coincide on the spectrum with those of the pigment. Fluorescent and LED lighting has discontinuous spectral composition, as shown in Figure 5.28. Because daylight and light from a 'black body' – e.g. tungsten or halogen – have a continuous spectrum, these can be regarded as a colour-rendering yardstick against which other illuminants can be measured.

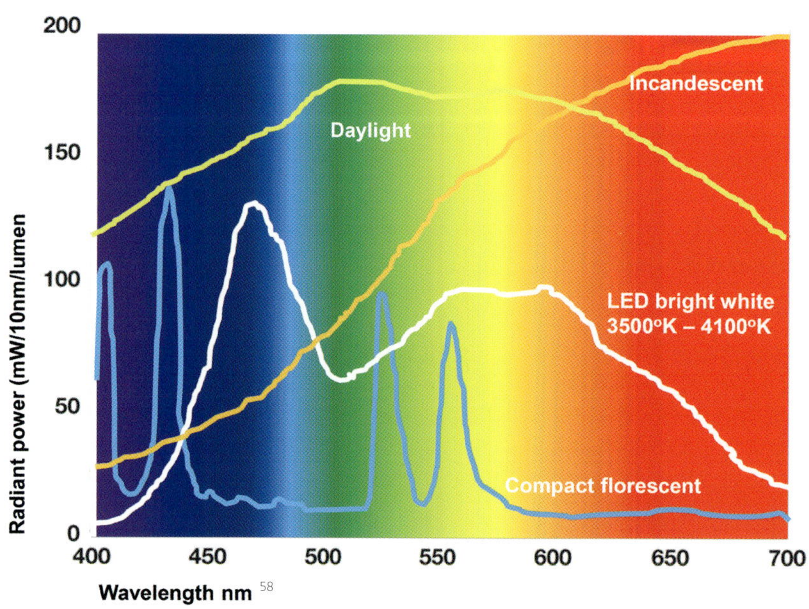

Figure 5.28: The spectra of various light sources.

It has been established that good colour rendering is desirable for visual comfort, and actually improves visual acuity.[59] Early in their development, both fluorescent and LED suffered from very poor colour rendering due to their strongly discontinuous spectrum. Later developments in both technologies introduced phosphors, which in effect filled in the gaps in the spectrum. Now, both lamp types are obtainable with good colour rendering and with a range of colour temperatures. Colour rendering is indicated by the Colour Rendering Index (CRI). Values above 85% are good and will be almost indistinguishable from full-spectrum light such as daylight.

Spatial distribution of light

Lighting is seen as a key component of aesthetic design, partly in the way it reveals the architecture (very much in the way that architects interpret daylight) and partly in the luminaire design itself. This can preoccupy designers and distract them from the light-distribution function of a luminaire in terms of the direction and intensity of light in the working area and, more importantly, the distribution of luminance on ceiling, walls and floor. Because our eyes sense luminance rather than illuminance, the reflectance of the room surfaces is also an issue.

Two extreme cases can be imagined (see Figure 5.29). Room A has a dark ceiling with recessed downlighters and a dark floor. Room B has a light ceiling with suspended translucent-glass globe diffusers, which have an upward light-output ratio of 50%, and a lighter floor. Both deliver 250 lux at the reference point. However, the perception of these rooms will be entirely different.

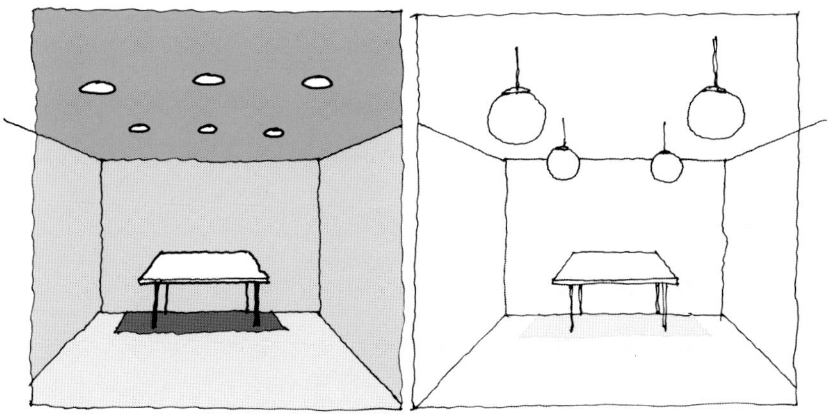

Room A Room B

Figure 5.29: The two rooms have the same illuminance at the reference point but have very different luminance distribution, and will be very differently perceived by the occupant.

Room A will have no direct light on the ceiling and only very little reflected back from the floor. The illumination at the reference point will be highly directional and therefore cast harsh shadows. The occupant's 'rest view' – i.e. a horizontal view not looking at the task – will be of the low-luminance walls. The lighting environment will have a very nocturnal feel, reminiscent of moonlight.

Room B will have bright walls, an even brighter diffuser surface and ceiling immediately above, and softer shadows. It will resemble the lighting distribution of outdoor daylight when the sun's disc is clearly visible through thin cloud.

We are very sensitive to the light distribution of a space. This is exploited when shop windows are lit, when actors are lit onstage and when recreational spaces such as nightclubs are designed. Clearly, the human brain has become used to a lighting environment that relates to a natural source – i.e. the sky – and the time of day.

Much research has been carried out for offices, looking at the preference of room appearance in relation to the illuminance distribution.[60] Its conclusion was that, in the daytime, people prefer a mixture of diffuse and directional light, to give soft shadows, and to have wall luminance relatively high. The researchers do not relate this to any natural cause. However, we propose here that lighting that supplements daylight should be introduced into rooms, which leads to a luminance distribution similar to that from daylight.

Recommendations for room lighting

Where artificial lighting is used to supplement daylight, it makes sense to use a source that is of high colour temperature – either bright white or daylight. This light will appear to be a similar colour to the daylight, and have similar circadian activity. It should not be used in the evening, when it will both appear ill matched to the ambient light and have the wrong circadian action. Luminaires should be chosen to introduce semi-diffuse light in a way that illuminates the walls and ceiling.

Separate light sources should be used for evening and night illumination, with lower colour temperatures. There is more freedom of choice of luminaire than for daytime illumination: directional for tasks such as reading, and diffuse for general ambient light. Light sources can be lower in position – e.g. table lamps – walls and ceiling can be of lower luminance and, except for task lighting, the overall illuminance can be lower than in daytime since it is not 'competing' with high levels of daylight close to the window.

Another possibility is to have luminaires with functions that automatically change the colour temperature of the light as the day progresses. This process, known as *dynamic lighting*, does not necessarily mean that the spatial distribution is changed. It has been installed in a number of schools and hospitals, and manufacturers of the lighting equipment claim positive results.[61] In all cases, the principle of adaptive opportunity should be observed – with intuitive, over-ridable controls.

Lighting for the sleep period

Another important consideration is lighting for limited activity during the sleep period. It is not uncommon for older occupants to require one or more visits to the bathroom during the night. In such cases, it is very undesirable that the full lighting system be activated – both for the benefit of the woken occupant and any sleeping occupants. Bearing in mind that the woken person will be fully dark-adapted, only very low levels of illuminance will be necessary (< 10 lux) to safely navigate doors, steps, etc. This should be of low colour temperature – e.g. a pre-set dimmed tungsten lamp, or an LED of appropriate power and colour temperature – and introduced as diffuse light, avoiding a view of the high-luminance source.

A similar approach, minimising the waking stimulus of bright light, could be of benefit in families with infants and young children.

Lighting controls

Conventional hard-wired light switching allows simple on/off and, in some cases, dimmer control of each luminaire or groups. Lighting 'scenes' – i.e. the creation of specific light distributions would be effected manually. Recently, digital and wireless technology have combined to make addressable luminaires. This means that provided power is delivered to the luminaire, all switching and dimming can be achieved from a hand-held controller. Furthermore, pre-set lighting scenes can be programmed and selected easily. This approach also offers the facility of repositioning luminaires more easily, allowing an occupant to tailor the room lighting to their own preference.

These systems can also offer light-management functions – i.e. lighting being switched off or dimmed if no presence is detected, or if ambient light from daylight is above a certain threshold.

A more recent development has been variable-colour lamps, also controllable wirelessly. In principle, this avoids having to duplicate light sources for day and night conditions. However, it does not address the issue of spatial light distribution.

Outdoor lighting

External lighting in the residential setting is mainly used for security, outdoor recreation and for decorative purposes. Manufacturers provide specialist advice, but here we draw attention to issues that may affect health and wellbeing.

It is obviously important that lighting does not cause annoyance to the occupier themselves or their neighbours. This can be achieved by careful consideration of lines of sight, and the use of shading or light baffles. Light should not fall on windows, even if normally curtained, since at night partial light transmission by curtains or blinds may still be sufficient to cause annoyance. This is particularly important when the light for security purposes is under movement-detection control – i.e. it can be activated by many false alarms, such as innocent passers-by. Better control avoiding false alarms may be achieved by separating the movement detector from the luminaire.

Where lighting is used for safety – e.g. enabling people to negotiate steps and obstructions – careful consideration should be given to glare, either from the luminaire itself or from over-illumination. It is little help to illuminate a single step so brightly that the person is left with poor vision to negotiate another, less well-illuminated hazard.

There are negative aspects to outdoor lighting. Urban skies are now so light polluted that clear views of our stars and planets are no longer possible. Light pollution also impacts on wildlife – creating false dawns, thereby disturbing animals' circadian clocks. Outdoor lighting also uses energy. It undoubtedly contributes to safety and security in many circumstances but it should be used sparingly, avoiding over-illumination. Where spaces are infrequently occupied, and/or at times when occupation is unlikely, switching by presence detection or time clock should be used. Luminaire design should be prevent wasteful light escaping upwards to the sky.

Light pollution from neighbours' outdoor lighting is now controlled by the Clean Neighbourhoods and Environment Act, section 102, and can be classed as a statutory nuisance under criminal law.

Key points

- Modern light sources such as fluorescent and LED are available in blue-rich, red-rich and intermediate spectral distributions. Blue-rich is compatible with daylight and is circadian-active – i.e. inducing wakefulness. Red-rich is also circadian-active, inducing melatonin production in preparation for sleep.
- Colour rendering is the ability of light to reveal true colours. Lighting should have a colour rendering index (CRI) of at least 85%.
- Artificial lighting used in the daytime should be of high colour temperature and have a spatial distribution (direction and illuminance) resembling daylight.
- Lighting controls can be used to switch and/or dim combinations of luminaires to suit time of day, and to prevent wasteful illumination of unoccupied rooms.
- Night-time (sleep-period) activities should be provided for, with very low-illuminance low-colour-temperature lighting.
- Outdoor lighting should be used sparingly, with care taken not to annoy neighbours and wastefully illuminate the sky. Avoid glare and over-illumination.

Internal noise control

The principles of noise control have been presented in some detail in 'Noise and its control by the envelope and structure' (chapter 4), but those applications mainly considered external noise sources. Noise generated within the home can also present problems.

A classic problem is amplified sound often generated by teenage children, much to the dismay of parents. The solution can be found with noise containment, using the principles of structural-noise reduction. The role of well-fitting doors, mass and rigidity, and acoustic absorption in reducing reverberation has been dealt with in principle, and can be applied here in practice. There is a reciprocal advantage in constructing rooms with good sound reduction; noise generated by adult life is kept from disturbing infants and young children.

A much simpler and less costly solution to the specific problem of amplified music is the use of headphones – however this is not under the control of the designer.

Noise management using these principles will undoubtedly increase costs, and there may be other objections to the principle of acoustic isolation. For example, it may be held that overhearing young children is part of parental supervision. However, doors can always be left open or, using the well-developed technology of baby alarms, one-way audible communication can be set up – giving the best of both worlds.

There is a growing move away from the use of carpet in residential buildings in favour of timber, stone, plastic-laminate and vinyl surfaces. Whilst probably being a positive move for air quality, this can result in a noisy household – due to both the generation of impact noise and the increase in reverberation.

Impact noise can be lessened by the choice of resilient finishes to furniture feet, and reverberant sound reduced by heavy curtains and soft furnishings. Removable (and cleanable) floor coverings such as rugs reduce both.

PART THREE:

Designing healthy homes

Chapter 6

Design scenarios

Introduction

These design scenarios have been developed to illustrate how the ideas, methods and practices described in this book can be applied in realistic situations. They include a range of housing typologies: high- and low-density housing, refurbishment and new-build, single to multifamily buildings, and accommodation for the young and the old. In each of the eight scenarios, certain issues are dealt with in more detail than others; it is not intended that each one comprehensively demonstrate all of the issues.

We have explored various criteria and solutions, and none are offered that have not included compromise. This illustrates the difficulty of weighting different considerations, which leads to conflicting solutions but reveals the design challenges and opportunities presented to the architect.

The scenarios loosely follow an imagined design development with a series of proposals, critiques and responses.

The following design scenarios are presented.

1 Medium-density low-rise new-build housing development
2 High-density medium-rise new-build urban neighbourhood
3 Low-density detached new-build house design
4 Existing high-rise 1960–1980 urban neighbourhood
5 New-build prefabricated mass housing
6 Victorian terraced house
7 Conversion of 1970s' detached house to care home
8 New-build student accommodation

Medium density low-rise new-build housing development

The brief is to design a new edge-of-town housing estate of 1,200 homes at a density of 50 dwellings per hectare, consisting mainly of 3- and 4-bedroom market homes and 30% social housing, with some 1- and 2-bedroom apartments in 4- to 5-storey blocks. The location of the development is well situated: adjacent to the green belt, a new school, an existing supermarket and good public transport (a guided bus five minutes from the town centre, and a park-and-ride facility with local bus and coach station).

The initial proposal draws on nearby traditional village typologies and materials, resulting in an organic layout of mixed detached and semi-detached homes (Figure 6.1).

Communal outdoor space and exercise

Objection

There is a lack of communal public space to encourage social interaction, physical activity and play.

Responses

Increase the proportion of 3-storey homes and the number of floors of apartment buildings, and increase the terraced-housing typology. This raises the potential for physical activity within the home – i.e. the use of stairs – and releases open space for public use.

The density of development increases the likelihood of significant car ownership per hectare, potentially making streets and public spaces less effective as places for social and physical activities.

Incorporate off-road car parking in the form of carports and garages; design streets as 'home zones'; introduce landscaping and traffic-calming measures to ensure that the streets prioritise pedestrian use and safety. Ensure that bicycle parking is incorporated and more easily accessible – i.e. near the front of the homes – than car parking, to nudge occupants to use bicycles as a healthier form of transport.

Objection
The incorporation of on-plot car parking risks increasing plot sizes and undermines the public-space benefits of the terrace building typology.

Response
Prioritise public space over private space by limiting the size of private gardens. The social-wellbeing benefits of good public space outweigh the benefits of large private gardens. Incorporate balconies and terraces to increase private outdoor space, access to fresh air and views.

Landscaping design and diversity

Objection
The public space is larger but potentially anonymous, with little variety for different kinds of activities under different weather and seasonal conditions.

Response
Design public space in such a way as to create diverse conditions (rather than one large main

Figure 6.1: Commercial (developer-led) housing layout of part of the proposed development.

area) including space for playgrounds; dog walking (separated from playgrounds); hard-landscaped spaces for walking, cycling and playing (notably when the grass is wet); and tree-planted space with benches for summer shade, allotments, wildlife areas, etc. Use landscaping and planting to enhance these qualities and diversities.

Distributing public green spaces throughout a development can result in more homes having access to views and access to, and control over, green space. As a result of the greater spacing between houses due to distributed public space, sunlight and daylight availability can also be increased – although this may be offset by the increased number of storeys of housing and apartment buildings.

Where possible, use public-art funds to create and enhance a stronger sense of identity of public spaces and add interest to encourage users to stop, engage, observe, think, etc. This can be achieved by incorporating landscape art or interactive objects, and wildlife-interpretation boards, in public spaces.

The masterplan for the Accordia housing development in Cambridge by Feilden Clegg Bradley Studios (Figure 6.2) exemplifies key ideas

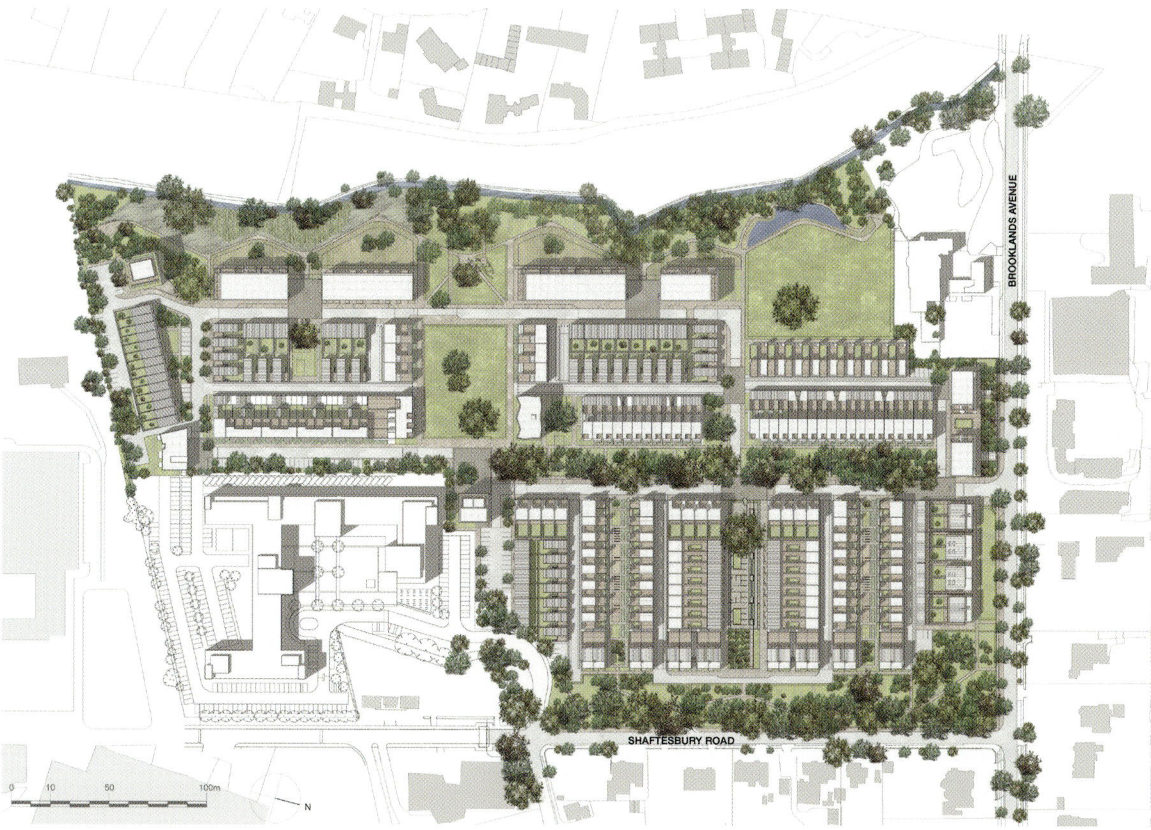

Figure 6.2: The masterplan for the Accordia housing neighbourhood exemplifies the use of distributed and diverse outdoor spaces to provide access to, views of and engagement with nature and neighbours that improve wellbeing.

of using housing form and diverse public-realm landscape strategies to improve wellbeing.[1]

Daylight and sunlight access, privacy and views

Objection

A greater number of storeys to the housing and apartments may reduce access to daylight, sunlight and views.

Response

The use of terraced housing instead of detached or semi-detached houses can partly help to offset reduced access to light and view, by increasing available open space and increasing average spacing between dwellings.

A three-storey home can incorporate a living space on the first floor to improve daylight and views. The knock-on health benefit is that occupants may use stairs more frequently between kitchen/ dining and the living room, and it might nudge them to prepare food and eat together in the dining area rather than in front of the TV. The challenge is to overcome privacy concerns due to views from upper-floor living rooms. This may be counteracted by careful design of the low portion of windows or views down from such windows – using, for example, balcony balustrades that obscure a downward view (although the compromise is that this also reduces the view of the occupants' own garden) but maintaining distant views of the landscape and sky.

Larger windows and bay-window seats can offer better views out to the neighbourhood and green space, and create settings with more daylight (circadian light) and potentially access to fresh air (if openable to a garden or green space) (Figure 6.3). Again, addressing privacy is important and can include ensuring that bay windows exploit more distant or oblique views and do not face directly onto near neighbours. Incorporating shutters and acoustic-ventilation louvres adds further control over fresh air, sunlight, noise, privacy, etc.

Figure 6.3: Bay-window seat offering daylight and views to the oocupant from an upper storey to a distant landscape.

High-density medium-rise new-build urban neighbourhood

The project is the redevelopment of 1.7 hectares of ex-industrial and commercial land in a medium-sized industrial city in the English Midlands. The brief called for low-to medium-cost apartments suitable for first-time buyers.

The site (Figure 6.4) lies to the north of mixed commercial/residential medium- to high-rise urban fabric (4–10 storeys), and south of a busy urban highway, a shallow and broken belt of 20th-century suburban housing, then open countryside with fine distant views of hills and moors.

The initial concept (Figure 6.5a) is for a series of 4-storey 25 m x 25 m blocks grouped around a court that provides secure over-looked outdoor space, protected from traffic noise and pollution, and landscaped or hard-play areas in alternative

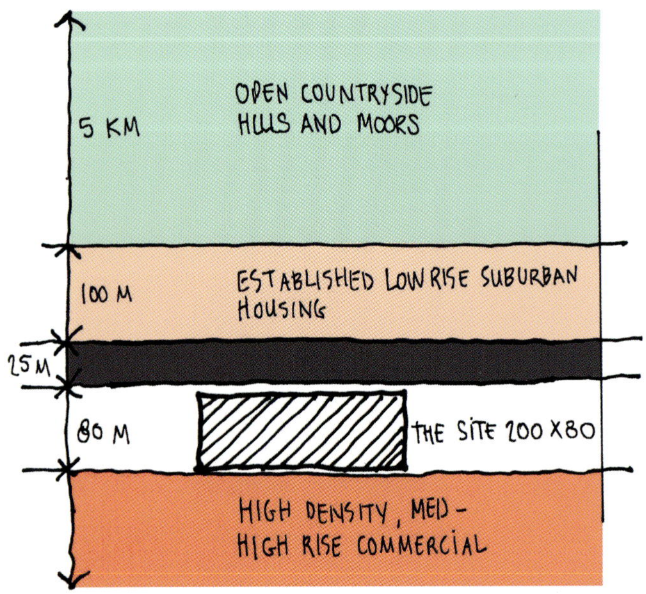

Figure 6.4: Site for high-density medium-rise neighbourhood development.

courts. The blocks would be separated from the road by a landscaped area 30 m deep. There would be a mix of single-storey flats and 2-storey maisonettes. The road-facing (north) façades would be acoustically defensive, with sealed windows and acoustic fresh-air vents. Internal layouts would place bedrooms facing the court to allow window opening.

Courtyard access to sky and view

Objections

The internal courts are too small (width-to-height ratio 4:5) to provide adequate SVF to ground-floor accommodation. Alternating play area and quiet area not likely to match needs of residents. 75% of accommodation gets no distant views.

Response

Enlarge courts and include both play and landscaped areas in all courts.

Objections

Enlarged courts still do not provide distant views (Figure 6.5b).

Response

Courts enlarged by opening at corners (Figure 6.5c).

Traffic noise and pollution

Objections

Openings provide conduits for traffic noise and pollution.

Response

Create profiled 3–4 m noise screen at roadside with landscaped bank facing housing. Blocks can now move closer to road, occupying the whole site (Figure 6.5d). The upper three storeys in roadside blocks (A) will have unobstructed distant views to the north. Ground floors will have close-up views of landscaped area, which will be sunlit during summer and at certain times

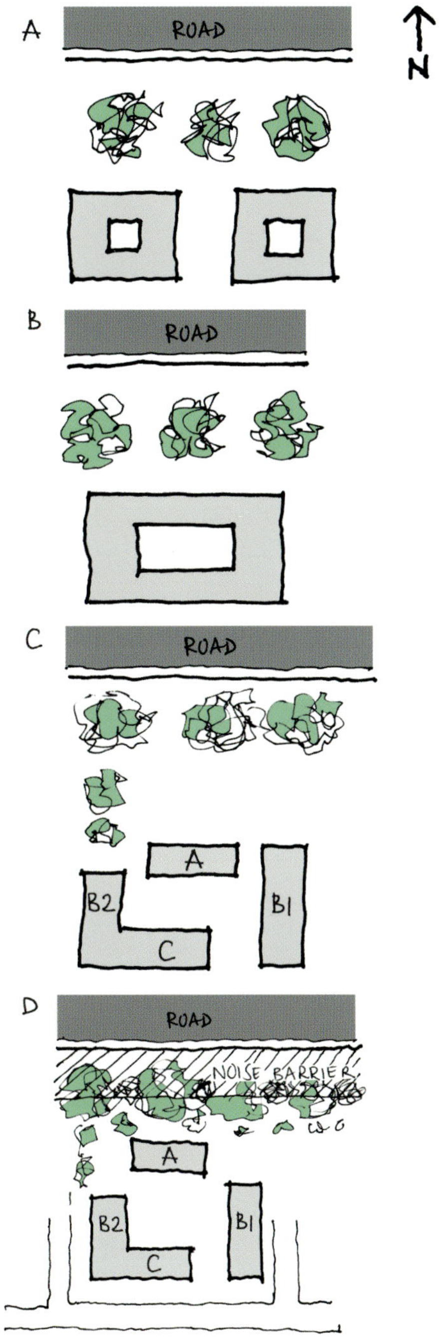

Figure 6.5: Sketch options for high-density medium-rise neighbourhood development.

of the day in all seasons, via openings in the blocks (Figure 6.6).

Ground and first floor will be maisonettes (duplex), allowing direct access to the ground for each household, for children needing supervision, and possibly small private gardens.

In roadside block A, floors 3 and 4 to have balconies designed to give acoustic screening from road, which is otherwise in line of sight from upper storeys. This will allow window opening for summer ventilation.

Other blocks (B) and (C) will have distant views via gaps in blocks.

Circadian issues

Objection

Solar access to block C is compromised by high-rise building to the south of the site, and in winter most of the court landscape will be in shadow. Blocks B1 and B2 have east-west orientation, giving morning and evening sun except in winter. B1 has the evening insolation from the quiet landscape area and morning from the noisier service area. This gives a pro-circadian location for bedrooms on the east side, with lounge/living area on the west. Block A has north-south orientation with some overshadowing to the lower floors and poor winter morning insolation, but good midday access.

Similar orientation for B2 would result in the living area facing the service area.

Response

Wall finishes on all court-facing blocks should be light (>50% reflectivity) and hard ground finishes should also be reflective. Dark evergreen planting, such as holm oak and cypresses, should be avoided or kept small. Adopt anti-circadian designation with lounge facing east into the

Figure 6.6: Landscaped earth-bank noise barrier screens upper storeys whilst providing sunlit landscape views to ground floor.

court. Maisonettes have at least one bedroom east-facing to the court.

Units that have poor early-morning solar access (A and C) should have kitchen/dining area in sunlit south-facing room. Possibly, ground floor in block C should be re-assigned to commercial and service use where heavily overshadowed by existing buildings.

All windows should be at least 20% of the floor area they illuminate – and for east, south and west orientations be provided with deployable shading.

Access to nature

Objection

There is poor access to countryside area to the north. Single-storey flats on second and third floors are isolated from immediate ground-level landscaping.

Response

Provide 'green bridge' across urban highway, continuing a 'green corridor' through suburban housing and through the new development to the commercial area. Get financial support from businesses to provide the bridge – this could bring benefit to employees, linking the town centre to recreational area via walking/cycling link.

Place interpretation boards (*'What you are likely to see'*) in commercial district, and in shared circulation areas of housing, to 'advertise' green link to countryside.

All second- and third-floor flats can have 'micro terraces' – small projecting areas adjacent to windows that can be safely accessed from inside, where plants and/or bird food can be placed. Projecting grids can be provided to catch dropped pots or containers to protect people below.

Low-density detached new-build house design

The accommodation brief for his project is: 180 sqm, 4 bedrooms, 3 bathrooms, 1 garage. The proposal is for a contemporary-designed energy-efficient family home (Figure 6.7). The location is near Kettering, Northamptonshire, on a semi-rural brownfield site with no traffic-related air-quality or noise concerns.

The main room on the ground floor is the living/dining room with a large open 'American style' kitchen and open fireplace as a focal point. Large windows allow good light penetration. There is a study room, a guest bathroom, storage room and an integral garage. On the first floor, there are four spacious bedrooms and a main bathroom. The master bedroom has its own en-suite bathroom.

The proposed construction consists of load-bearing concrete-blockwork cavity walls with

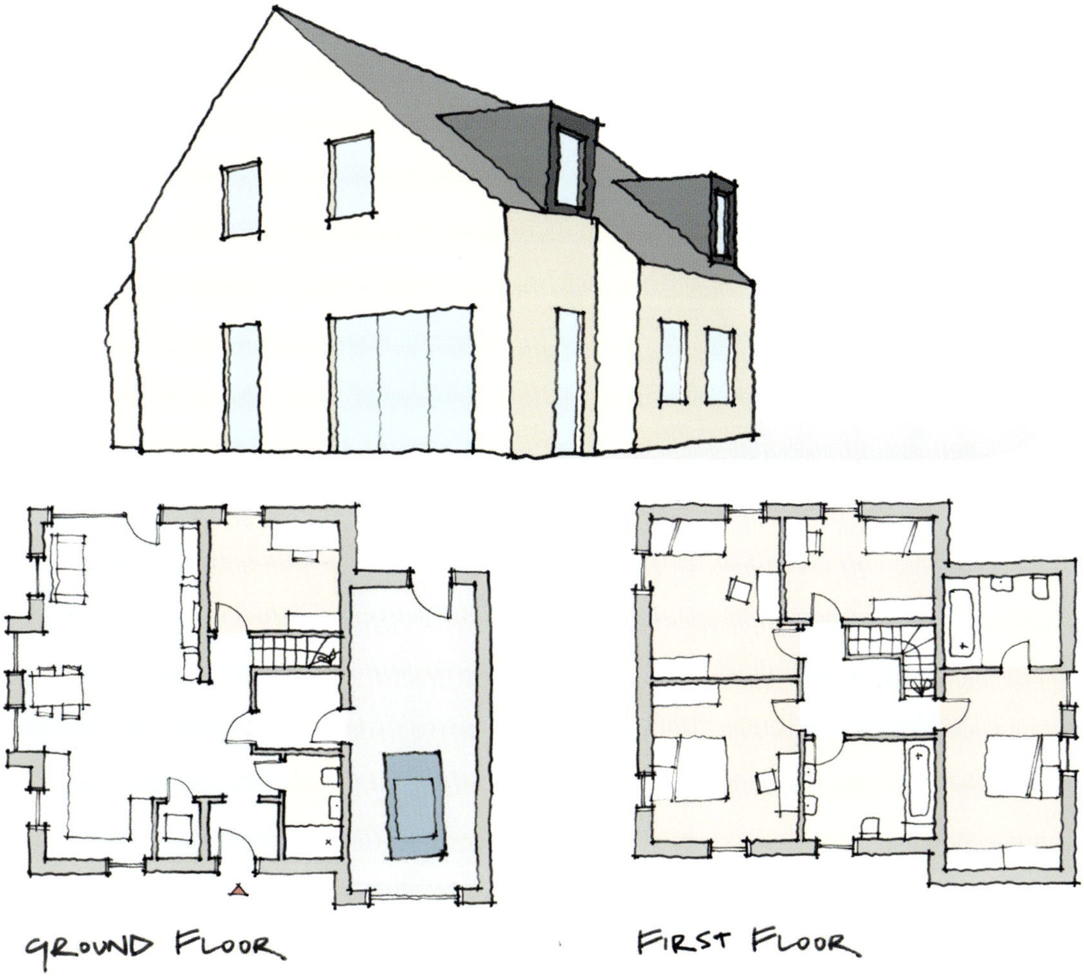

Figure 6.7: Typical contemporary design for a four-bed detached family house.

roof structure with slate tiles. The design meets strict energy-efficiency standards – equivalent to PassivHaus – and uses mechanical ventilation heat recovery (MVHR).

Radon and indoor air quality

Objection

On further site investigation, supported by Public Health England's affected-area maps,[2] the site near Kettering is found to be contaminated and radon is found to be present in the ground. Radon, a radioactive gas, is a leading cause of lung cancer after smoking. If not dealt with, it can build up in the home – particularly if the MVHR and other extract fans create a slight negative pressure that draws the gas in via the ground floor and foundations.

Response

The design of the ground-floor construction is adjusted to incorporate a gravel bed or permeable layer, covered with plastic sheeting to act as an impermeable layer. This layer is then linked via a well-sealed connection though the plastic sheet to a ventilation pipe that runs up through the house and ventilates out above the roof. An accessible in-line fan is used to guarantee adequate airflow. To improve the protection, the original raised timber floor is replaced by an in-situ concrete floor over the plastic sheet (Figure 6.8). This prompts a reconsideration of ecological materials used elsewhere in the construction, resulting in the decision to use a timber-frame construction (reduced embodied energy) on concrete foundations and ground floor.

Objection

The domestic indoor environment can dominate personal pollution exposure due to the large proportion of time on average spent at home and the presence of pollutants related to domestic activities. The awareness of radon in this project has prompted general concerns about the effects of ventilation on indoor air quality and health.

Response

There is evidence that mechanically ventilated buildings have a higher average ventilation rate than naturally ventilated ones. This is particularly the case (and important) in winter, when MVHR also brings energy savings. Thus, MVHR was proposed. Furthermore, low-level permanent ventilation extract is applied in bathrooms, kitchens and utility/storage spaces, with local higher rates automatically triggered when humidity (and the associated risk of damp/mould) and contaminants (from cooking for example) are sensed.

Timber construction can make it easier to achieve lower U-values and limit cold bridging (and damp/mould risk), but needs to be sealed

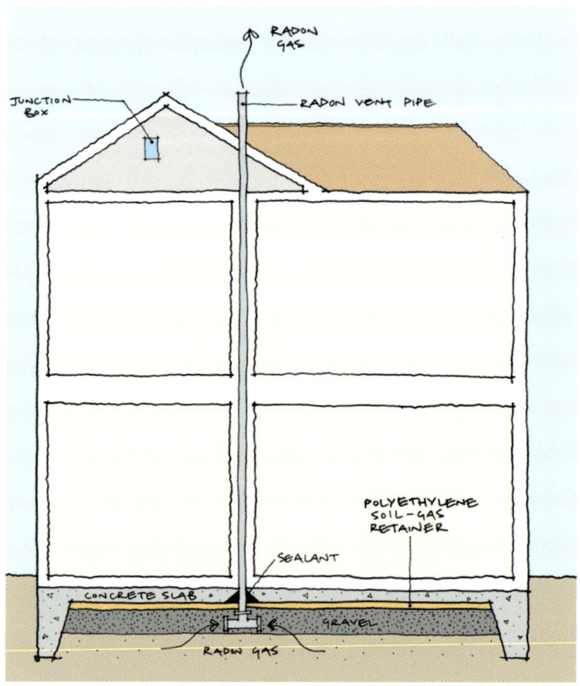

Figure 6.8: Design of a radon-ventilation system.

with an airtight membrane to avoid interstitial condensation, minimise ventilation losses and for the MVHR to be most effective.

MVHR and open windows

Objection

Beyond radon, moisture and mould control, there are other building-construction considerations that have health implications and add to the potential pollution exposure of occupants.

Response

Materials specifications are drawn up to limit chemical content, off-gassing, VOCs and dust. Avoidance of emissions due to formaldehyde, particleboard, phthalates in plastics, paint, carpets, furnishings, etc. are important in reducing asthma, allergies and respiratory problems. A cross-laminated timber (CLT) form of construction is proposed with a natural finish – i.e. no varnish or stain – and carpets are omitted in favour of a tiled ground floor and timber upper-floor finishes.

To limit allergens from household dust mites, a regime of occasional high ventilation rates and cold temperatures are recommended. This requires auxiliary heating beyond MVHR and PassivHaus principles to bring the house back up to temperature. The added complexity, space and cost of MVHR is not considered justified if auxiliary heating is also required (underfloor on the ground floor and skirting radiators for the upper floors), although extract fans with sensors are still adopted. The slightly reduced energy performance is offset by passive and active solar systems.

Thermal mass core

Objection

Lightweight construction – e.g. timber/CLT – has a greater risk of overheating, notably on upper floors, which can increase heat stress and related health consequences (particularly for more vulnerable, less mobile occupants such as the old and very young). In the context of climate change, these risks are likely to increase.

Response

The concrete slab and tiled floor finish provide thermal mass to the ground floor. A masonry structural core is proposed to rise through the centre of the house to provide access to thermal mass at the other levels. To encourage natural ventilation even on still days, the house is now three storeys tall and the stairwell creates a passive stack with automatic ventilation openings at the top (with rain sensors and manual override). The three-storey dwelling also has health benefits because it encourages greater physical activity through stair use (Figure 6.9). Windows are designed to incorporate ventilation louvres (with insect screens and insulated shutters). Thus, secure night-time ventilation can be provided in the summer without ingress of insects (likely to become an increasing problem with climate change). In winter mode, the insulated shutters can be closed and only trickle ventilation is available (Figure 6.10).

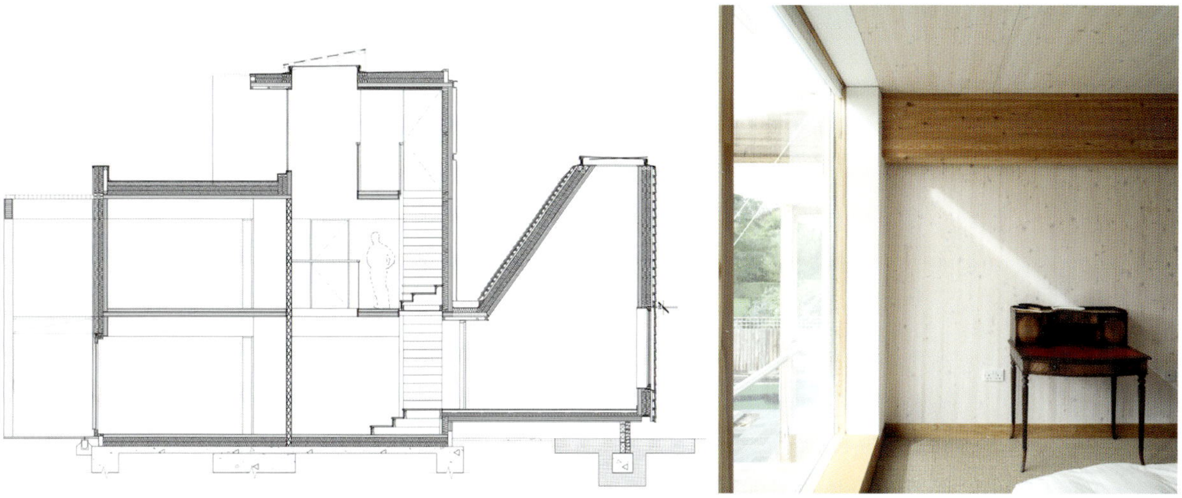

Figure 6.9: This design for a private house in Cambridge by Mole Architects demonstrates the health-design principles outlined above: section through stairwell/stack (top), natural timber interior (left) and view of stairwell with masonry wall (right).

Figure 6.10: Louvred (left) or perforated (middle) acoustically lined opening with insulated interior shutter (right) adjacent to window to enable noise reduction and secure (night-time) ventilation.

Existing high-rise 1960–1980 urban neighbourhood

Background

The existing tower block was built in 1967 using the Large Panel System (LPS). This consists of storey-height prefabricated concrete panels, in which window and door openings have already been made and fitted. They were usually dry-lined with 25 mm of expanded-polystyrene (EPS) insulation and plasterboard set on battens fixed to the concrete. Assembly relied on bolted connections stabilised with grout. Structural stability relied solely on the panels and the integrity of the junctions.

In 1968, a small gas explosion on the 18th floor of Ronan Point, a 22-storey residential building in south London, generated sufficient pressure to dislodge a wall panel, which led to the progressive collapse of the whole corner of the building, killing four and injuring 17. The result of this was the removal of gas installations from all buildings of this construction type, and their replacement with electric storage heaters.

The project here is for the refurbishment of an 11-storey LPS tower with two 2-bedroom and two 1-bedroom flats on each floor (Figure 6.11). The block has its main glazed façades facing east and west. It has its original internal insulation (25 mm EPS) and electric storage heaters. Low-income families, couples and individuals, some elderly, occupy the block. There is widespread evidence of fuel poverty: mould growth and condensation, supplementary heating by portable electric and liquid-petroleum gas (LPG) appliances, and ventilators blocked up.

Consultation with the occupants reveals that the major perceived issues are with the cost of heating and its poor control, and the resulting condensation. They are generally unaware of the cause of condensation, often claiming that it is rain penetration, and equally unaware of the need for ventilation.

We are mainly concerned here with providing an affordable but healthy thermal environment, but will also consider the problems of social isolation.

Insulation

External insulation and renewal of single-glazed windows with UPVC double-glazed low-E (low-emissivity) units is proposed. The insulation can be installed without displacing the occupants, and the windows can be replaced with temporary disturbance only. There would be a massive improvement in heating effectiveness. Furthermore, cold bridging due to the internal insulation and the panel construction, the site of much of the mould growth, would be eliminated. But though comfort will improve, fuel costs will not decrease in proportion to the reduced heat loss since the occupants are cost limited already. There would also be frequent complaints of overheating, with wasteful window opening the

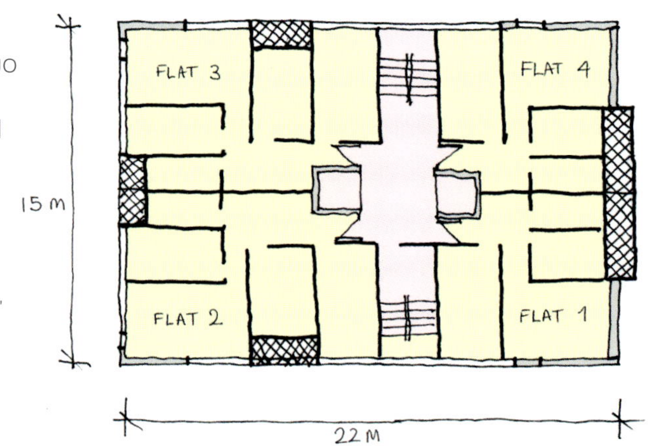

Figure 6.11: Typical floor plan for an existing (1967) 11-storey residential tower of Large Panel System construction.

only recourse to the occupant. Thus, the insulation must be accompanied by major upgrading of the heating system and controls (see below).

Note: During the preparation of this book, the tragic fire at Grenfell Tower, west London, occurred with large loss of life. Fire spread from a local source on the 4th floor via the external cladding and insulation, up and into rooms on all 21 floors above. This thwarted the building's compartmentalisation strategy, rendering the 'stay put' policy for the occupants tragically inappropriate. Lack of sprinklers, fire-resistant doors and poor escape routes were also cited as a contributory cause of loss of life. This incident will have a major impact on all technical aspects of the refurbishment of high-rise residential towers, with respect to fire safety. This issue is beyond the scope of this book.

Heating, controls and ventilation

A central gas boiler is installed in a new annex at ground-floor level. Each flat has its own wireless heat meter, and also a minimum temperature function that provides unmetered heat to prevent temperatures dropping below a pre-set value. *(Analysis may show that this is unnecessary due to the low thermal resistance of the separating walls and floors).* Control would be a simple fixed day/night programme with a central 'Warmer' and 'Off' manual-override control, both reverting to original programme on the next cycle. Panel radiators would have pre-set non-adjustable thermostats to prevent overheating from solar or other gains.

Objection

The above actions result in a large reduction in energy costs and a further increase in comfort. But air quality is still a problem, with high humidities sufficient to promote mould growth in poorly ventilated areas such as clothing storage. Domestic activities such as clothes drying, washing and cooking, together with respiration, can result in 5 to 10 kg of water-vapour production per day. Other pollutants from cleaning products, plastic-based materials, cooking and possibly smoking will also be present. New window frames and overcladding are likely to decrease the infiltration rate further, making deliberate ventilation essential.

Response

Extract ventilation must be installed in kitchens and bathrooms. The kitchen control point should be humidity and particulates (smoke). The bathroom control point should be humidity and presence.

Objection

An ingress route for air should be provided. Fire regulations will probably prevent this from being via the internal stairwell and landing. The problem of fixed trickle vents in window frames is that they lead to over-ventilation (and draughts) at times of high winds and large temperature differences. Occupants often then seal them up.

Response

Incorporate self-balancing vents. These contain a damper that closes at higher pressure difference, thereby stabilising the flow of air.

Objection

Summer conditions will normally require openable windows to obtain high rates of ventilation. Their openable area should be at least 10% of the floor area - and preferably in two or more parts of the wall to improve air circulation.

Response

Provide for securing the window at various degrees of opening, including one or two settings to obtain a minimal ventilation rate, as well as security against entry - particularly important on the ground floor.

Objection
Pivoted windows when open may present acoustically reflecting surfaces, which can increase the transmission of sound from a certain direction. For example, a horizontal centre-pivot window, when opened bottom-outwards, will tend to reflect noise generated at ground level into a room. Side-hung windows hinged so they face each other can transmit sound from room to room.

Response
Ensure that side-hung windows hinge to face away from each other and avoid horizontally pivoted windows.

Circadian issues

The floor plan shows all habitable rooms on the perimeter with generous glazing areas. Thus, the building will be well daylit.

Objection
The west-facing bedrooms will receive low-angle summer sun late into the evening, making them brightly lit (anti-circadian) and possibly overheated. East-facing bedrooms will receive morning sun (pro-circadian), but in the summer this may be much too early.

Response
The necessary response in both situations is effective shading and openable windows (see above). By *effective*, we mean *controllable* to near blackout but also *deployable*, so that the shading device can be removed from the glazed aperture completely – allowing uninterrupted views and maximum daylight penetration at times of poor sky luminance. Note that this does not mean dark-coloured blinds, but that that are of low translucence. Blinds are obtainable that are white to reflected light but have zero light transmission.

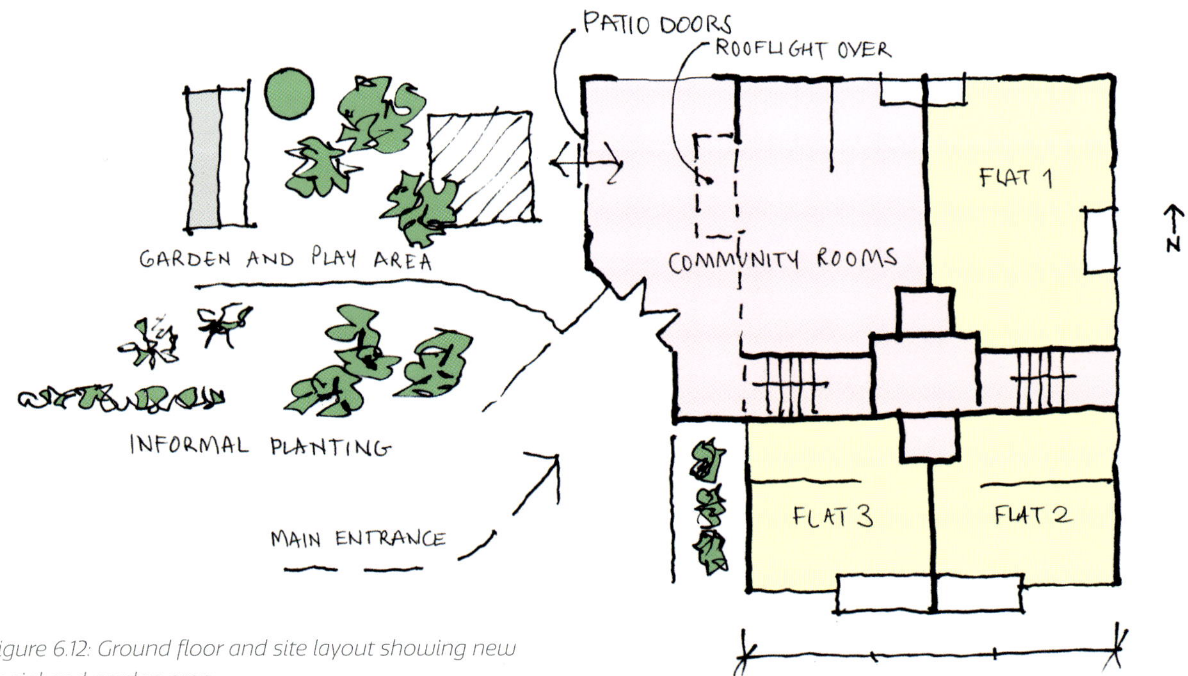

Figure 6.12: Ground floor and site layout showing new social and garden area.

Social and nature isolation

Social isolation is more likely to arise for elderly residents living alone. It may also occur for younger adults who may be unemployed, or single parents with young children.

Objection

There currently seems to be no social space within or adjacent to the building, and whilst there may be community facilities in the area, for people of poor mobility and motivation these may be ineffective.

Response

It is proposed that one ground-floor flat be taken to form a remodelled entrance, including social spaces and a toddlers' play area. Immediately adjacent, accessible via openable patio doors, an enclosed garden area with raised beds and permanent seating is established. Stairwells and landings are to be refurbished with indoor planting, decorative art and comfortable seating. The entrance will improve privacy for the existing ground-floor tenants (Figure 6.12). An alternative (if structural alteration to ground-floor flat is unfeasible) is to build a separate but linked 'pod' for these functions.

As part of the window replacement, 'micro balconies' can be installed for the safe placing of plants and bird food.

New-build prefabricated mass housing

The aim of this project is to construct housing quickly, cheaply and sustainably. The approach is to design modular homes using off-site manufacturing whilst enabling a degree of customisation to suit different households. The attraction of off-site construction is that homes can be made to high performance standards – with increased airtightness and thermal insulation. In this scenario, there will be more focus on the construction and interior opportunities for health and wellbeing, and less discussion of site and context.

The initial concept proposes a timber-framed construction with Structural Insulated Panels (SIPs), clad in brick or fibre-cement panels, and metal roofing. Floor-to-ceiling heights and window sizes are relatively generous compared to the norm. The stairs and plumbing (kitchens and bathrooms) are fixed but the remaining plan layout is bespoke in terms of open-plan versus subdivision – e.g. number and size of bedrooms – and a choice of upstairs versus downstairs living room (Figure 6.13).

Toxic construction materials

Objection

The choice of construction, materials and finishes are focused on delivering a quick, lightweight and energy-efficient home. Construction-choice implications for health with regard to the indoor air quality should be addressed because it is these indoor conditions to which occupants are exposed for typically 70% of the time. Harmful indoor chemical emissions can include:

- formaldehyde from
  - particleboard
  - plywood
  - foam insulation
  - carpet underlay

    - furniture
- phthalates from
    - plastic materials
    - vinyl flooring
    - flame-retardant chemicals
    - adhesives
    - sealants
    - paint and coatings.

Such materials can be 'off-gassing' chemicals for a period of months or more. Furthermore, carpets and soft furnishing can act as a reservoir for household dust mites and allergens, as well as a cocktail of household chemicals. These can have a direct effect on asthma, allergy and respiratory symptoms; some are carcinogenic, and others can cause developmental problems in children.

Response

The best way to reduce the exposure of occupants to indoor pollutants is to control them at source. Thus, firstly, in the specification of materials, avoid those mentioned above that are associated with chemical off-gassing or biological pollution within the home, and specify non- or low-emitting adhesives and alternatives. LEED, the WELL Building Standard and International Living Future Institute's 'Red List' all provide advice on this – as seen in 'Pollution sources – occupants, contents, activities, outdoor air' in chapter 5.

Secondly, although significantly less effective, where it is not possible to eliminate all potentially harmful chemicals, ensure that such materials are sufficiently ventilated before being enclosed within construction components or sealed in to the prefabricated modules and 'shrink-wrapped' at the factory. Finally, a last resort is to ventilate the house freely and extensively before occupation whilst running the heating system, as well as facilitating easy free ventilation and cleaning during occupation.

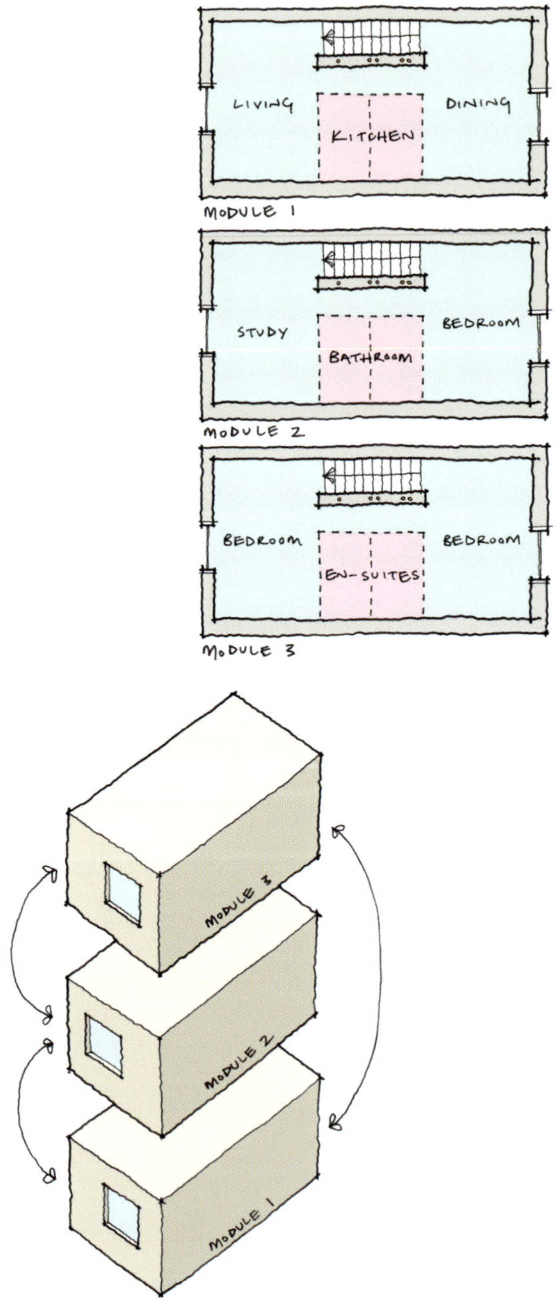

Figure 6.13: Sketches for a 3-storey home and various plan configurations – e.g. the living room could be on the ground, first or top floor.

These last two approaches lengthen the time from prefabrication to occupation, which in turn challenges one of the key advantages of off-site construction – speed. Nevertheless, with prefabrication it is essential that chemical pollutants are not encapsulated within the modules and delivered to site.

Overheating

Objection

A lack of thermal mass – in combination with relatively large windows, and high levels of insulation and airtightness – raises concerns about overheating. Apart from thermal discomfort, this can cause heat stress, particularly for vulnerable occupants such as very young children and the elderly.

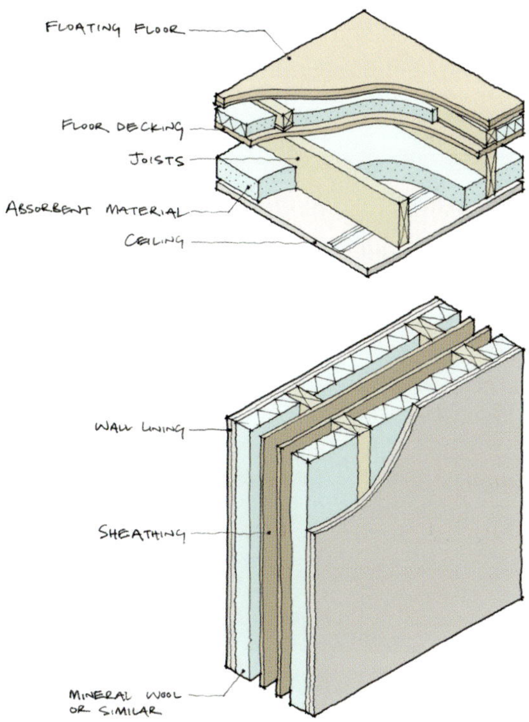

Figure 6.14: Separating floor (top) and wall (bottom) details designed to meet higher acoustic isolation than conventional internal floors and partitions.

Response

The principles of avoiding overheating have been described elsewhere, and include

- integrating shading components – external
- large openable sections of windows – unconstrained by restrictors, yet safe
- louvred sections – ensuring cross-ventilation and security, and incorporating insect screens.

Construction specification can increase thermal-mass effect without significantly increasing construction mass – e.g. dense plasterboard, phase-change plasterboard,[3] tiled floors, etc. Furthermore, systems can be considered: a ground duct for passively pre-cooling incoming air (ensuring that radon is kept out!) combined with a stack or mechanical ventilation strategy, or ground-source cooling combined with underfloor coils.

Indoor noise

Objection

With a timber floor construction and open-plan circulation and arrangements, acoustic isolation between spaces becomes an issue. Noise disturbance between rooms can be physically and psychologically stressful, potentially affecting sleep and limiting the acceptability of activities such as making or listening to music, or boisterous child play.

Response

Additional attention to sound-insulation strategies for floors, partitions and doors should be considered. This will require that internal partitions and floors be considered more as separating walls and floors between dwellings in their detailing (Figure 6.14).

Victorian terraced house

The house, built in around 1900, had an original layout and room use as shown in Figure 6.15. It has undergone a number of alterations since the 1980s.

It has solid walls, a concrete ground floor and now a mix of single and double glazing. Subsequent modifications up to the present are shown in Figure 6.16:

- front parlour and kitchen knocked into one large living room
- pantry extended with single-storey lean-to over outside paved area, to form larger separate kitchen
- back bedroom converted to bathroom and toilet
- middle bedroom divided to form two small single bedrooms
- outside toilet rebuilt to have indoor access
- (not illustrated) front wall removed and garden paved for parking.

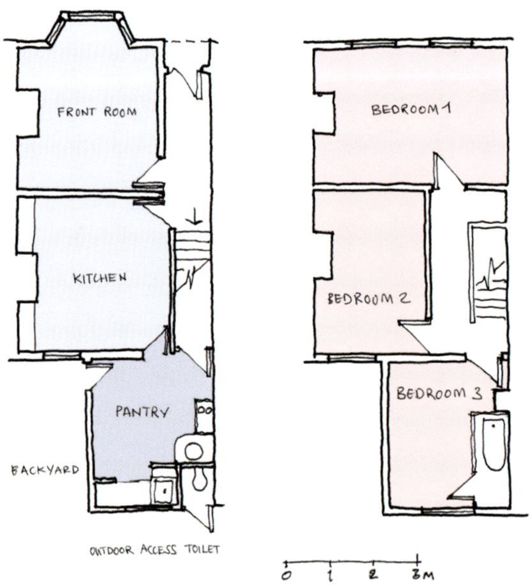

Figure 6.15: The original floor layout as built in 1900.

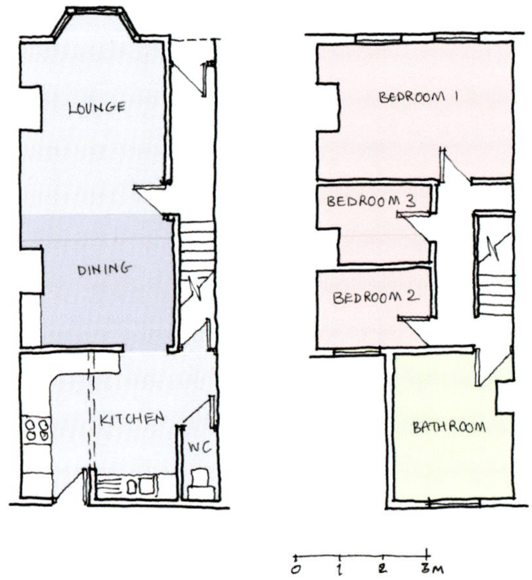

Figure 6.16: Floor layout as found in 2017 after alterations from 1960 to 1980.

The building is heated with an old gas boiler serving radiators without TRVs on a single circuit. The original walls are solid and uninsulated; the walls of the extension are cavity but uninsulated. The loft is insulated with about 50 mm of fibreglass. The front downstairs windows were replaced with UPVC double-glazed units about 20 years ago; these are now distorted and poorly fitting with weather strips missing. The remaining windows are single-glazed timber, showing signs of decay. There is also much evidence of condensation and mould growth in the bathroom, kitchen and some bedrooms.

Energy conservation

The initial proposal is mainly concerned with energy conservation and takes a conventional approach:

- increase loft insulation to 300 mm
- replace all windows with UPVC thermal-break low-E double glazing

- insulate external walls (front and back) with proprietary external system
- install condensing gas boiler
- replace radiators on separate circuits, with 2-zone programmer (ground and bedrooms)

Reassignment of room use

Objection

The existing modifications have created serious environmental problems. The scullery extension has obstructed the window at the deep end of the enlarged living room, resulting in very poor daylighting and subsequent underuse of this space. This has been exacerbated by dark finishes and recessed downlighters, resulting in very poor ceiling illumination and generally low luminance. Access to the downstairs toilet direct from the kitchen is also unacceptable.

The division of the middle bedroom in order to convert the back bedroom to a bathroom has resulted in one of the small bedrooms having no window (other than an obscure glass window onto the landing). There is also poor acoustic separation via a lightweight partition.

Response

Before energy-conservation measures are implemented, it is proposed to remodel the building and reassign the floor plan as follows (Figure 6.17):

- remove extension
- relocate ground-floor toilet to top of stairs
- enlarge kitchen area
- reassign rear of living room to dining/family room
- reassign old bathroom back to third bedroom with high-level glazing onto lightwell
- reassign old internal bedroom to bathroom by constructing new soundproof partition (see 'Noise and its control by the envelope and structure' in chapter 4) and installing extractor via duct in old flue
- install rooflight and well through roof space over stairway.

Pollution, noise and summer ventilation

The location of the building, facing a busy urban road, creates hazards relating to

- pollution
- noise.

Objection

The former is a problem all year and could possibly be mitigated by sealing the street façade and providing cleaner air from the garden area via ducts. These could run in the ceiling void, bypassing the kitchen and using the stairwell as a

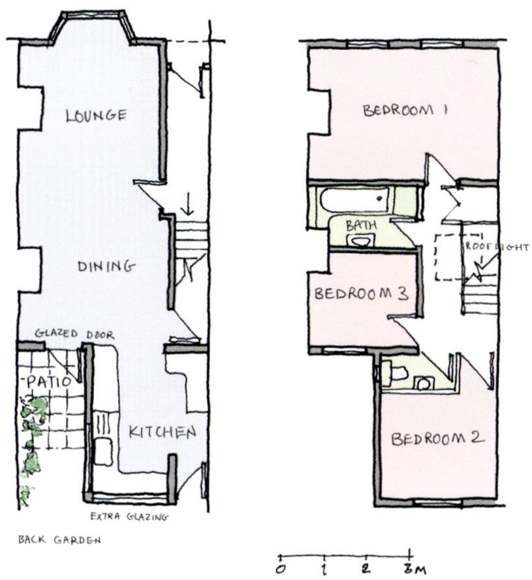

Figure 6.17: Proposed refurbishment restores daylight and garden views to dining area and third bedroom.

plenum – and could thus serve most other rooms. It is likely that the supply duct would require a fan, controlled by a combined CO_2/presence detector/timer. Extract points already exist at the kitchen and bathroom. Alternatively, it would be simpler to rely on the openable rooflight, although this would tend to act as an extract in winter. A passive extract stack (for quietness) could be installed in the front bedroom.

Response

The noise problem is more acute in the summer when, during warm weather, normal window opening on the façade facing the road would be inappropriate due to both pollution and noise. This is particularly the case in the front bedroom at night time. For a solution (if partial), effective shading in the daytime would ensure that the bedroom did not become overheated. Then, access via a louvred door or panel to the landing, which could be refreshed at high rates via the openable rooflight, would provide relatively clean make-up air, whilst the passive stack could provide extract. A low-speed ceiling fan could provide air movement. Calculations would be needed to check on flow rates and duct sizes.

Noise exposure to the street façade at ground-floor level could be reduced by building a solid masonry wall with heavy timber gates up to 1 m high. This would screen high-frequency engine and tyre noise, and could reduce the migration of particulate pollution. But it may also compromise ease of car parking.

Views of nature

Objection

The view from the kitchen and dining area is very precious since it this the only private garden space seen from downstairs. The view from the original window position in the back wall of the main terrace is very restricted.

Response

We have proposed repositioning and enlarging the old pantry window to allow views from the kitchen and dining area of a much wider area of garden. However, the window from the dining area is still important for ventilation and daylight; it follows that the new paved area needs to have surfaces of high reflectance – i.e. the ground, fence and wall – and vegetation should not be allowed to obstruct the glazing.

Lighting the deep plan

Objection

The deep-plan building suffers from large daylight illumination gradients across its floor plan. This can be alleviated by supplementary artificial lighting in the dimmer areas. However, both the colour and the spatial distribution should be compatible with daylight.

Response

We propose using luminaires with a high upward light ratio and high-reflectance upper walls and ceilings, together with day/night colour-temperature-change lamps. An alternative would be separate luminaries and sources for the daylight-supplement function and normal night-time illumination.

If the rooflight is adopted, an interesting possibility would be to open up the staircase to the dining area. If the rear wall of the stairwell had light reflectance, this would make a significant contribution to daylighting in the dining area and to the general impression of spaciousness. It would, however, have a negative effect on noise containment between ground and first floor.

Alternative relationship to the street

The original design was produced at a time when streets had a very different function. In 1900, mobility would be achieved almost exclusively on foot and by horse-drawn vehicles. Local pollution by CO and NOx would have been absent. Although coal burning would have resulted in particulates and SO_2, these would not have been concentrated at road level. Furthermore, due to the prevalence of pedestrian activity the street was a far more social space than it is today, serving as a meeting place and play space. This social importance was reflected in the design of the terrace frontage and the designation of the rooms behind.

Objection

This poses a dilemma. Do we turn our back on the street, recognising it as a source of noise and pollution, and serving only as a transport conduit? Or do we try to improve the street environment, repair its social function and maintain the response of the house?

Response

Many new-build designs would adopt the former strategy. Locating service areas such as cooking and bathrooms on the street side of the building with extract ventilation would create a prevailing ventilation direction from the less-polluted garden. Locating the main living areas away from the road would ensure good visual and physical access to the garden, and relative freedom from traffic noise.

The latter strategy i.e. restoring the original relationship to the street could realistically only be considered as part of a general improvement scheme that would involve multiple houses – indeed, the majority of the street. It would also necessitate a major reduction or elimination of traffic in order to reduce the impact of pollution, noise and physical danger. This compromise might be achieved by the designation of a shared pedestrian-priority street, which allows vehicular access (at greatly reduced speed) and parking but no through traffic.

Conversion of 1970s' detached house to care home

This scenario is based on the conversion and extension of a 1970s' four-bedroom bungalow to a residential care home (Figure 6.18). It stands on a narrow rural site overlooking farmland.

CARE HOME 1

Figure 6.18: The original room layout prior to conversion.

The declared aim is to maintain the small-scale domestic atmosphere, although subsequent extensions will eventually increase the number of bed spaces to 20. We have assumed that the extension is done in two stages.

Residents range from healthy, largely self-sufficient and ambulant, to bedridden – and display various degrees of dementia.

Three issues receive special attention:

1. access to nature
2. natural and artificial lighting
3. ventilation.

Phase 1: The first extension

Accommodation is increased by the addition of six en-suite bedrooms on the west side of the building. A new larger kitchen and storage area is formed in the area hitherto used as a garage and the old kitchen is reassigned as a garden lounge – particularly for the use of residents in the west wing, giving the views of the garden and distant countryside (Figure 6.19).

'Vertical gardening'

Objection

Views from the new bedrooms 5, 6 and 7 are very limited. Bedrooms 1, 8 and 9 look out onto the approach and car-parking area.

Response

Bedrooms 5, 6 and 7's positions are compensated for, to some extent, by 'vertical gardening', including planters and bird feeders on the boundary fence. The car-parking area is landscaped with raised beds to keep vehicles from parking close to the windows. Bedrooms 2, 3 and 4 have prime views of the garden and distant landscape. The sitting room now incorporates a dining area.

General access to the garden is provided via the garden room (formerly dining), which still provides an alternative dining area, and the lounge. A north- and east-facing open veranda is installed to provide a 'soft edge': an area protected from rain suitable for gardening activities. This has a glazed roof in order to have minimum impact on daylight levels in the adjoining rooms.

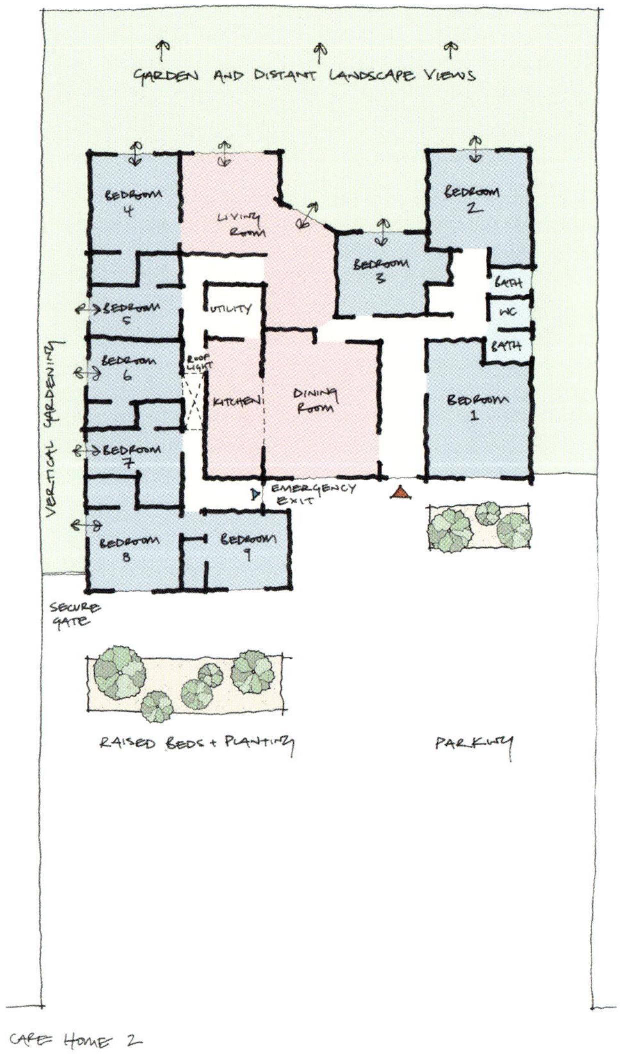

Figure 6.19: Phase 1 proposal.

Phase 2: The new wing and courtyard

This is a major enlargement in the form of a block lying across almost the width of the site.

Objection

Initially the major objection was that this cut off the existing accommodation from the garden and distant countryside views.

Response

In mitigation, the courtyard formed by the new block is landscaped with a semi-mature tree and raised beds, both decorative and suitable for vegetable gardening projects by the residents (Figure 6.20).

'Green' roof and garden sheds

The new block has a low- and double-pitched 'green' roof based on sedum species and including dry-habitat plants such as Rudbeckia, Achillea and ornamental grasses. This gives a three-dimensional quality to the roof covering, which can be seen from the courtyard, lounge and some bedrooms. The roof is a double-inclined structural insulated panel (SIP) with sloping ceiling soffits and the opportunity for rooflights along the ridge over the corridor, and high-level openable windows in the bedrooms (Figure 6.21).

An L-shaped partly glazed open veranda provides a covered link between the new lounge and a range of three 'sheds' – a gardening shed, a project shed (small workshop), and a 'theme' shed. The last-named can be set up with props and decoration to adopt a theme such as a beach hut, a '60s café or a log cabin. Or it can be used to house a temporary exhibition of, for example, antique cooking utensils, or an exhibition art and craft made by the residents or a local artist.

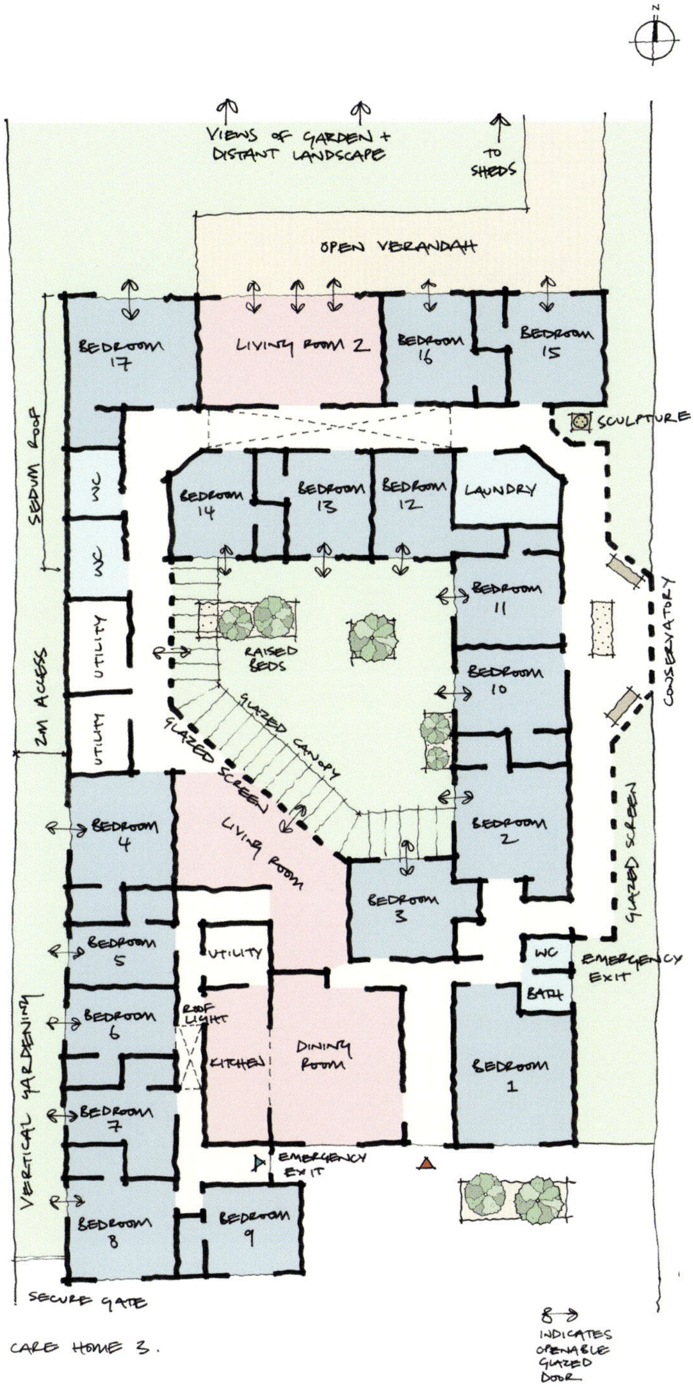

Figure 6.20: Phase 2 proposal.

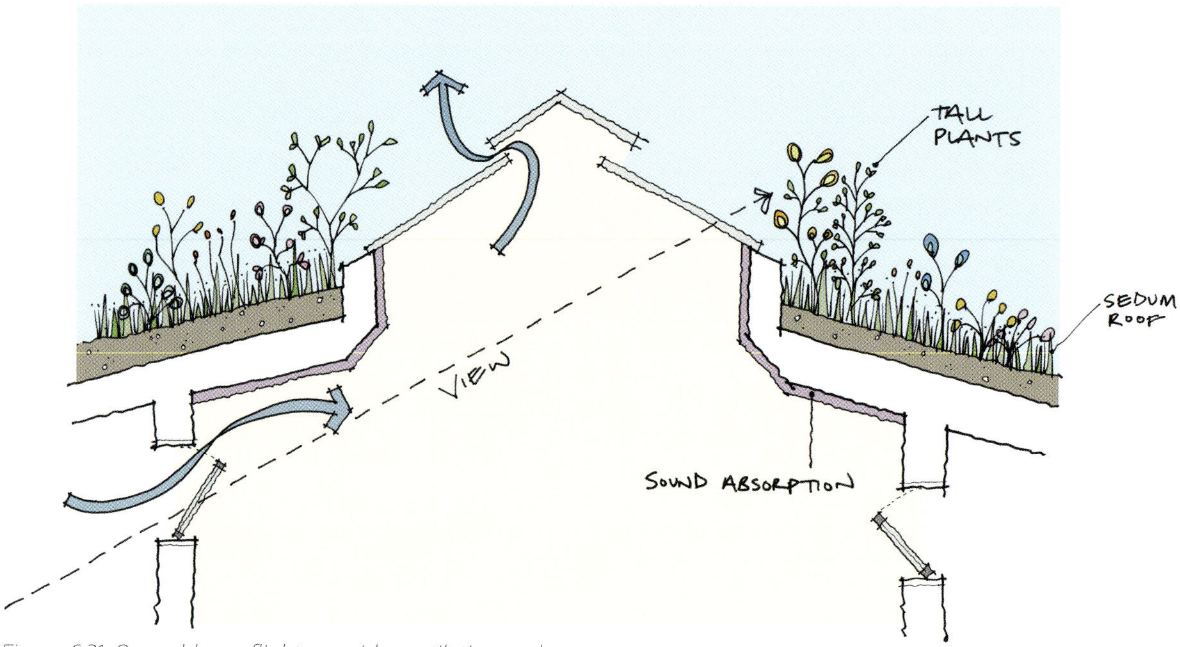

Figure 6.21: Openable rooflights provide ventilation and daylight via the 'green' roof.

The remaining garden is laid out in three ways:

1. a risk-free area that all residents could occupy
2. an 'adventure' area with changes of level, branched paths with dead-ends and secluded seating areas
3. a horticulture area for more advanced gardening activities than can be supported by the courtyard beds.

There is also an area of mown grass where residents could exercise by ball play or movement routines.

Natural and artificial lighting

Objection

All bedrooms should be well daylit – but this means designed to have a minimum 2% DF, which for the UK means that 300 lux is exceeded for roughly 50% of daylight hours.

Response

Curtains and/or blinds will allow this high level to be reduced. Blinds or other shading devices must be provided in rooms that receive significant solar gain.

Objection

The new bedrooms 4–7 are located close to the western boundary, facing the neighbouring development and the boundary fence.

Response

When specifying glazing area and position, the effect of such external obstructions must be taken into account. This could be achieved by clerestory windows or skylights, but it is very important that some of the glazed area has a low window sill in order that a seated occupant has a good view of the outside. Glazed doors to outside are provided in almost all rooms to give both pedestrian access and visual access to ground level, and to increase the level of connection with the outside.

Roof lighting, through openable roof windows via light wells, can be used to punctuate corridors and create focus to help orientate the residents. They can also be used to ventilate the corridor (see discussion on 'Natural ventilation').

Objection

Due to the orientation of the site (long axis north–south), sunlight penetration is difficult. Both lounges face north.

Response

There may be the possibility of sunlight penetration via rooflights. Glazing can be increased on the south-facing sitting room and the area close to the perimeter treated as a sun room. Blinds should be provided here so that all or part can be shaded at any one time.

For artificial lighting there should be separate daytime and night-time sources, changing from high colour temperature (containing more blue light that supports alertness) to low (for the evening and night time). In situations where daylight access is not achievable, electrical illumination during the daytime should be high in both circulation spaces and sitting areas, at around 450 lux – i.e. typically 50% higher ambient light levels for the elderly compared with those appropriate for younger people. Night-time illumination in the bedrooms should be dimmable down to 10 lux, to allow sufficient vision for safe movement without sleep disturbance.

Natural ventilation: windows and rooflights

A target is to maintain good air quality, primarily by natural ventilation. This can be achieved in warmer seasons by openable windows. Windows should have a range of opening between 'trickle' – i.e. sufficient to open the weather seal a few millimetres – to wide, or as wide as security considerations permit.

Corridors can be ventilated via rooflights, provided the roof structure makes the construction of ceiling wells practicable (Figure 6.21). The free ventilation of corridors means that the adjoining rooms can have cross-ventilation, either via a purpose-provided grille or an open door. (Residents frequently leave doors open in order to maintain social contact.)

Objection

Cold weather presents a different challenge, although higher temperature differences make the stack effect more reliable. If room-to-corridor vents are provided, corridors and rooflights could still serve a role in this. However, this will need careful control to ensure that the opening areas respond to different weather conditions (wind-speed and temperature).

Response

Self-balancing vents (see 'The driving forces of natural ventilation' in chapter 5) should be considered. Another possibility would be a centralised tempered-air mechanical supply.

Areas of high pollution, kitchen, bathrooms, toilets etc, would, of course, be ventilated by mechanical extract, making sure that there is a suitable air-supply route that does not interfere with other ventilation patterns. There must also be automatic control, sensing occupancy and with a time delay.

Understanding the building

Objection

Residents of a building such as this will have widely ranging levels of environmental awareness and ability to respond.

Response
It is essential that staff are well trained in the operation of windows, blinds, fans, lighting, heating controls, etc. and understand what the objectives are.

Objection
Residents might additionally be unable to articulate their environmental needs explicitly.

Response
The residents themselves should be informed about how the building works. For example, if one complains that the east corridor feels cold in winter (because of cold radiation from the glass), it should explained that they should consider it as 'a bit like outside', and that if they walk on to the conservatory, even when it's not sunny, there will be a warm refuge (with warmed seats) there. Or, if the sun room is too hot for one resident but being enjoyed by others, to move to the shady part of the room (where the temperature will seem to be about 5°C lower) or pull some of the blinds down to create shade.

Complaints relating to comfort should not always be seen as a negative issue; they are evidence that the resident is still being stimulated by the environmental diversity on offer. They should be used to prompt explanation and conversation about the building.

Graphic signs positioned in various parts of the building and relating to the local environmental issues (such as those referred to above) could be helpful, and could serve to stimulate an interest and involvement in the building.

New-build student accommodation

Mental health

Objection
With 2.3 million students currently in higher education in the UK and a steady increase in full-time students, the demand for student housing is high. This coincides with increasing reports of student stress and mental-health challenges.[4] The need to create healthy student accommodation is more urgent than ever.

The commercial opportunity presented by student housing has been exploited since 2010. Unfortunately, poor-quality accommodation has been justified on the basis that students spend little time in their rooms – and then, they are mostly asleep. This is exemplified by the poorest designs attracting negative press coverage, announcing '"Prison-like" student housing wins Carbuncle Cup ...'[5] and comments such as:

'The majority of rooms lack adequate daylight, offer little privacy and a significant number ... have no direct view out at all... approved on appeal by the planning inspectorate on the grounds that students don't require the same quality of accommodation as the rest of society.[6]

Clearly, we are a long way from Denys Lasdun's iconic University of East Anglia student residences of 1962, the pedestrian-only linking galleries of which were said to have been planned on the now utopian-seeming dictum that 'a minute spent walking is a minute spent thinking' (Figure 6.22).

Response
It is important to recognise that students can spend a significant amount of time in their rooms, over a number of years. This means that effort needs to be invested in designing physical and social conditions that support health and wellbeing, and not relegate student

Figure 6.22: UEA student residences, Norwich (1962) by Denys Lasdun: elevated walkways.

accommodation to lesser standards and expectations than other dwellings. Student rooms are places of sleep, but also of study and social interaction. Design criteria should include the application of circadian design principles to encourage healthy light-dark cycles. Blackout and noise reduction to facilitate sleep, daylight to support wakefulness, and views to provide contact with the outside are important considerations in establishing a design brief.

Orientation

Objection

Typical design for student accommodation consists of narrow-frontage single-bedroom units with individual shower rooms, along single- or double-loaded corridors with a small, shared kitchen and living space for every 4 to 8 bedrooms (Figure 6.23). Sometimes conceived as prefabricated units, the same room configuration and façade design is used for all orientations and

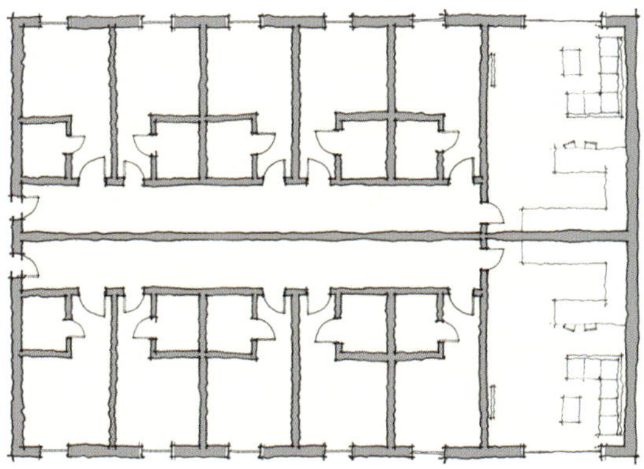

Figure 6.23: Typical student-accommodation layout.

on all floors, although variation in façade cladding is sometimes used to add variety.

Response

Rooms that are oriented north should be shallower than south-facing ones because the southern sky, even excluding direct sunlight, provides 60% more daylight than the northern. However, south-facing windows require external shading to limit overheating from solar gains in the summer. The advice is thus to orientate the majority of rooms towards the south, the view or away from significant noise sources and obstructions, depending on the context.

Social dimensions

Objection

Contrary to the condescending attitudes mentioned above, students need more than just a place to sleep. However, the push towards privacy, with individual bedrooms and en-suite bathrooms, has not helped to avoid isolation and depression. The focus of attention and resources on individual facilities has been drawn away from common spaces for socialising.

Response

The response should be to include a diversity of room types, more shared rooms, shared bathrooms and social spaces for larger groups of students in order to encourage interactions and foster a sense of community.

Commercial considerations have made en-suite designs common, but there is a resurgence of interest in more sociable arrangements. The potential reduction in size of individual spaces and the reduced servicing costs that result from not having en-suite bathrooms for every room

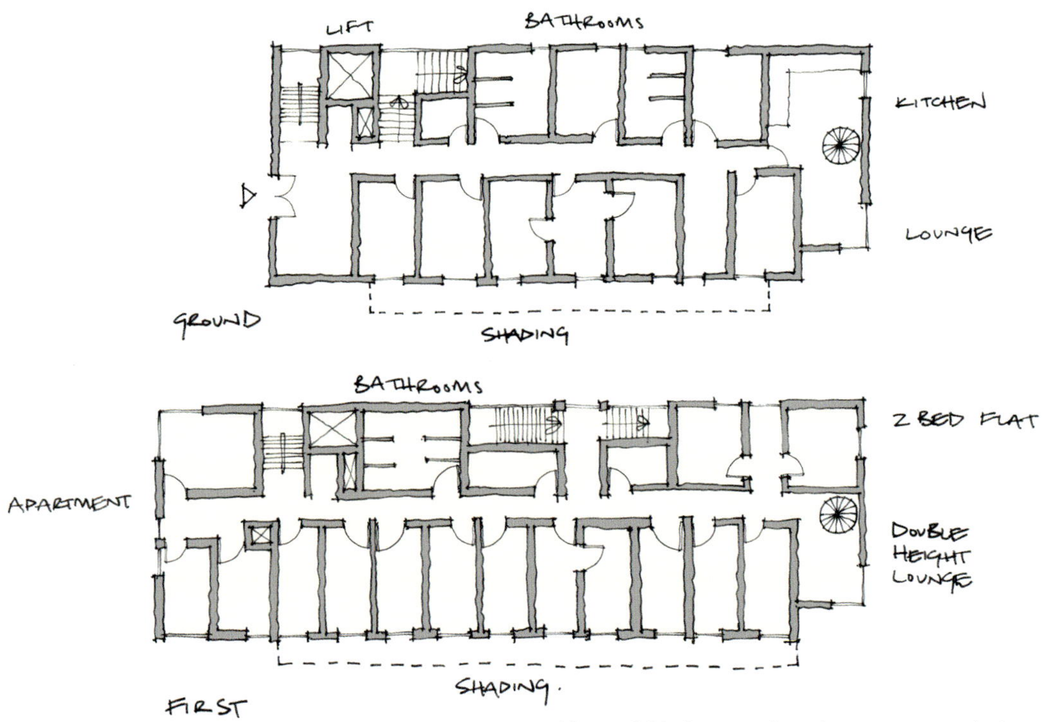

Figure 6.24: Proposed student accommodation.

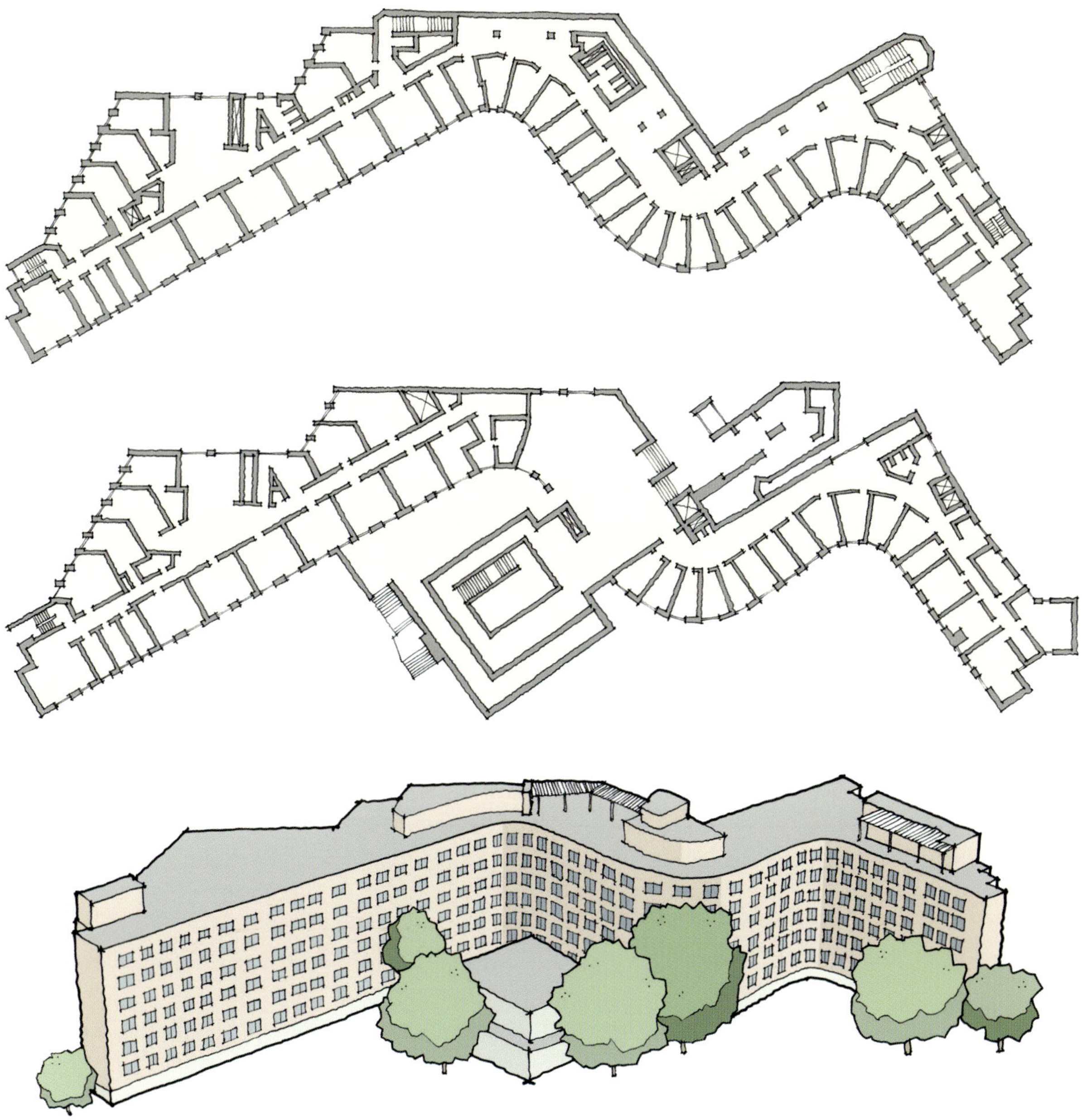

Figure 6.25: Aalto's Baker House (1946–48) at MIT is a worthy case study. Typical floors (middle) have a predominantly southeasterly sinuous façade of study bedrooms (of various sizes and configurations), overlooking the Charles River, with shared bathrooms and communal facilities on the northern side and on the ground floor (top).

means that more space and money can be allocated to communal areas.

In some situations, rooms with en-suite bathrooms can be more easily used for conferences – although this is not always the norm. Conference use does provide an opportunity to reduce costs for students and increase income for building owners, suggesting that some en-suite design might be appropriate in certain locations (and if conference facilities in or near the building can be accommodated).

Diversity

The proposed plan across two floors shows the majority of the roughly 20 study rooms facing south (with external shading devices), and bathrooms and circulation to the north (Figure 6.24). Room types vary from single to shared, including the possibility of a small number of apartments (for staff, visitors, married couples or young families). The vertical circulation links between the two floors is via a void in the communal spaces. A shared lounge is located to get the best views and light on the southeast corner, linking vertically to the adjacent floor and kitchen area. Fire routes are designed so as to minimise the subdivision of circulation spaces, and doors are held open magnetically to ensure maximum socialising in the communal space.

On the ground floor and basement, shared and semi-public facilities can be provided, such as a gym or sports facility, meeting or hot-desking areas, dining and café space, access to outdoor gardens, and more public commercial facilities. These provide opportunities for wider community connections, physical activities, contact with nature and healthy-eating options.

One can find many of the above characteristics in Alvar Aalto's design for student accommodation at the Massachusetts Institute of Technology (MIT) (Figure 6.25).

Chapter 7 Conclusions

Concern for human health in relation to the design of buildings is not new. The growth of biological science in the 19th century was frequently directed towards environmental aspects of housing and the workplace. Although economic and political pressures meant that the environment that many people occupied was far from healthy and safe, it might have been worse but for the efforts of people like Florence Nightingale.[1]

Heating, ventilation, air quality, the availability of daylight and the impact of noise were all issues upon which the Victorian scientist and architect had discourse. Yet more than a century later, we cannot claim to have understood these issues fully and finally – much less incorporated their implications into everyday architectural and environmental design.

Why the continuing debate?

Several factors have contributed to the continued debate. Firstly, society demands an ever-increasing standard of environmental convenience and health. For a school to send its children home early because the urban sky was so dark from pollution and the school's budget for lighting was running low would now be unthinkable. Similarly, for six households to share an outside toilet would be equally shocking. But in the 1890s, both school and urban sanitation, with all their limitations, were recognised as huge advances in the intellectual and physical health of the nation.

Secondly, new environmental stresses are being created by modern lifestyles. For example, although our cities may nowadays be rid of thousands of smoking chimneys, we now have PM2.5 and NOx to contend with from road traffic. Sleep disturbance from street occupants and traffic may not be new, but has grown with the proliferation of restaurants, bars, nightclubs and other entertainment serving the 24-hour urban culture. Aircraft noise now affects large numbers of households in the vicinity of airports. High-density high-rise living creates many new inter-neighbour issues such as a lack of acoustic and visual privacy, friction due to pet keeping and even car parking!

The interior environment is not only affected by a building's surroundings but also by new agents from within, often those to which our body has no evolved response. For example, modern carpets, paints, laminates, adhesives and cleaning materials all have serious effects on indoor air quality. Furthermore, our ability to monitor and understand our environment has increased immeasurably due to the development of affordable technology and scientific knowledge.

Finally, we are in a highly dynamic and transient situation. New hazards are around every corner – population growth; urban pollution; water and food security; drug resistance; energy security; climate change; and, just recently in terms of public awareness, oceanic micro-plastics. These problems continue to grow exponentially and time horizons shorten. It is not surprising, then, that the debate continues.

Future-proofing

How then, can we respond in our design of housing? How can we respond to existing factors, *and* pre-empt the future? Climate change introduces a whole new emphasis on designing for uncertainty. If it only brings a steady increase in temperature, we might be able to cope. But what of changes in seasonal amplitude and extreme events? What of changes in other meteorological parameters such as wind speed, precipitation, humidity and its impact on the already challenging urban environment in which an ever-increasing number of people live? It is often pointed out that the average life of a house exceeds the human lifespan – can we then, future-proof our homes not only for our contemporaries but for future generations?

The short answer must be that we cannot. But we can respond to *existing* scientific knowledge on health and wellbeing, *provided we can make the link between the scientific evidence and the design solution*.

'Good enough' design

In this book, we have tried to set out the various environmental factors that, according to the evidence, affect human health and wellbeing. We have also described the physics of these factors and how it can lead to the engineering of the building fabric. It would be nice, then, if based on this logical construct we could recommend an optimum solution that perfectly balances the values of all these factors.

But how do we weight the relative importance of, say, good daylight penetration deep into a living area against the risk of overheating and the transmission of traffic noise (not to mention less technically determined concerns of visual privacy versus a view out)? Wherever there is a compromise to be made, where the value of a parameter – in this case, glazing area – is pushed in the opposing direction, it would need far greater epidemiological study than we have to date to determine 'optimal' solutions. Furthermore, due to the temporal variability of the drivers – e.g. sunlight availability, traffic noise – an optimum may be fundamentally impossible.

So firstly, we have to resort to a fall-back position of *good enough* design – design that although not optimised in certain functions, performs reasonably well over a wide range of conditions. This *safe territory* design would, within predictable limits, not allow obvious failures – and would use the intuition developed by experience rather than a strictly mechanistic model to reach good compromises.

We are accustomed nowadays to justifying expenditure only when strictly evidence-based. Due to the complexity of the environment, and the heterogeneity of its occupants, it is difficult to provide hard evidence of benefit resulting from all the measures advocated here. However, many are cost-neutral, or nearly neutral – maybe costing a little more design time to think through the implications and chose the right option.

'Nudge' architecture

We also have to face the fact that there can be no guarantee that occupants will use a building in a way that maximises their own health and wellbeing – and even less, that of the planet. Education can help with this, but instruction manuals (and now, presumably, training apps) issued with new buildings have never been a resounding success and rarely survive beyond the first occupation.

But by making it just a bit easier and more rewarding to do the right thing, can design nudge people into environmentally healthy behaviour? Much thought goes into the ergonomic design and layout of an aircraft cockpit (including such details as making the lever controlling the landing gear actually look like the landing gear). This attention to intuitive design is meant to ensure that under conditions of stress – or even boredom – it is more likely that the pilot will do the right thing. Whilst the manifestation of such an approach to building design might be different, the same principles could be adopted.

Adaptive opportunity

Finally, we resort to that powerful solution – *adaptability*. At an evolutionary level, adaptability has been *the* major influence on the biological success of the human species. Paradoxically, during our recent exponential rise in population and sophistication, the need for adaptability in our built environment has been neglected. However, this has come at a great cost – in energy, in resources, and possibly in health and wellbeing.

Earlier in this book, we placed great value on adaptive opportunity in order to achieve comfort. We could extend this principle to provide not only immediate adaptive opportunities for occupants but also the ability of the structure and infrastructure to be adapted to meet changing conditions. Some of these, such as global warming, we see around the corner, but there are also others that we are yet to face.

In the early 1970s, the RIBA President Alex Gordon coined the phrase 'long life, loose fit'. As the energy crisis developed, this was extended to 'long life, loose fit, low energy'. Now, nearly a half century later, should we add '... and healthy'?

Maybe this would be an appropriate mantra for the uncertain future.

Figure 7.1: The availability of natural light and fresh air, combined with access to nature, are basic principles of design for sustainability and wellbeing.

Glossary of acronyms and key terms

adaptive comfort theory that humans can be comfortable over a wider range of thermal conditions than predicted by simple heat balance, provided they have the opportunity to make alterations to their dress, behaviour or the environment, which tends to restore heat balance

advection flow caused by a difference of pressure leading to the bulk flow of moist air

albedo reflectance of building surfaces

BRE Building Research Establishment

BREEAM BRE Environmental Assessment Method

BSI British Standards Institution

CABE Commission for Architecture and the Built Environment

CIE International Commission on Illumination (*Commission internationale de l'éclairage*)

CFD modelling modelling of fluid-flow problems using Computational Fluid Dynamics

CLT cross-laminated timber

CO carbon monoxide

CO_2 carbon dioxide

colour rendering the ability of light to reveal true colours

CRI Colour Rendering Index

Daylight Factor (DF) ratio of illuminance inside to unobstructed illuminance outside a room

dewpoint temperature at which air becomes saturated and condensation occurs

DHW domestic hot water

diffusion flow caused by the random motion of molecules

EEA European Environment Agency

FAIRMODE Forum for air-quality modelling in Europe

free running a building's 'free-running' temperature is that attained without the use of heating or cooling

GPs General Practitioners; another name for UK NHS doctors

heat balance heat loss through a building envelope exactly equals heat gain

heat-island effect temperature increase observed within a city when compared with the outlying rural area

heat recovery extraction of heat from outgoing vitiated air to preheat incoming fresh air

HQM Home Quality Mark

IAQ indoor air quality

interstitial refers to condensation that takes place within a structure

IPPC (United Nations) Intergovernmental Panel on Climate Change

ISO International Organization for Standardization

LST denotes low-surface-temperature devices

MV mechanical ventilation

MVHR mechanical ventilation with heat recovery

NHS (UK) National Health Service

NICE (UK) National Institute for Health and Clinical Excellence

NIHR (UK) National Institute for Health Research

NO_x oxides of nitrogen

NO_2 nitric oxide

NV natural ventilation

OPEC Organization of the Petroleum Exporting Countries

PACs polycyclic aromatic hydrocarbons

PM Particulate Matter

relative humidity the mass of water in the air divided by the mass of water if the air were saturated at the same temperature

reverberation time time taken after sound source cut off for reflected sound to become inaudible

SAD Seasonally Affective Disorder

SBS Sick Building Syndrome

sgf solar gain factor

SIP (prefabricated) structural insulated panel

sky view factor measure of amount of sky visible from a relevant point on a building

Sonnenbaulehre solar theory related to urban planning: it seeks, through the orientation and arrangement of buildings, to ensure increased access to sunlight and air flow in order to create healthy living conditions

SRI Sound Reduction Index: the difference in sound level as sound travels through a partition

STH Social and Therapeutic Horticulture

thermal admittance combination of thermal capacity and conductivity

TRV thermostatic radiator valve

UKGBC UK Green Building Council

US DHHS US Department of Health and Human Services

U-values thermal transmittance values (measured in W/m^2K)

VOCs Volatile Organic Compounds

WHO World Health Organization

WRAP Waste and Resources Action Programme

Further reading

Barton, H., Thompson, S., Burgess, S. and Grant, H. (eds). (2015). *The Routledge Handbook of Planning for Health and Well-Being*, Routledge: Abingdon.

Bluyssen, P. (2013). *The Healthy Indoor Environment*, Routledge: Abingdon.

Channon, B. (2018). *Happy by Design*, RIBA Publishing: London.

Day, C. (2015). *The Eco-Home Design Guide: Principles and practice for new-build and retrofit*, Green Books: Cambridge.

Dolland, T. (2018). *Designed to Perform – Energy Efficient Homes*, RIBA Publishing: London.

Heywood, H. (2018). *101 Rules of Thumb for Low Energy Architecture*, RIBA Publishing: London.

Hopfe, C.J. and McCleod, R.S. (2018). *Passive House Designer's Manual*, Routledge: Abingdon.

Levitt, D. and McCafferty, J. (2018). *Housing Design Handbook*, Routledge: Abingdon.

Littlefair, P. (2011). *Site Layout and Planning for Daylight and Sunlight*, BRE Press: Watford.

Meek, C. and van der Wymelenberg, K. (2015). *Daylight and Integrated Lighting Design*, Routledge: Abingdon.

Mills, S. (2018). *Fundamentals of Architectural Lighting*, Routledge.

Nicol, F., Humphreys, M. and Roaf, S. (2012). *Adaptive Thermal Comfort: Principles and Practice*, Routledge.

Park, J. and Porteus, J. (2018). *Age-Friendly Housing*, RIBA Publishing: London.

Phillips, D. (2004). *Daylighting: Natural light in architecture*, Architectural Press: Oxford

Ritchie, A. and Thomas, R. (eds) (2009). *Sustainable Urban Design: An environmental approach*, Taylor & Francis : Oxford

Stewart, J. and Lynch, Z. (2018). *Environmental Health and Housing*, Routledge.

Szokolay, S. (2014). *Introduction to Architectural Science: The basis of sustainable design*, Routledge.

Tragenza, P. and Wilson, M. (2011). *Daylight and Lighting Design*, Routledge.

Whitehead, R. (2009). *Residential Lighting*, Wiley : Chichester

Bibliography

Adams, A. (1996). *Architecture in the Family Way: Doctors, Houses, and Women, 1870-1900*, McGill-Queen's University Press: Montreal.

Age UK (2012). 'The cost of cold: Why we need to protect the health of older people in winter', Age UK: London.

Aked, J., Michaelson, J. and Steuer, N. (2010). *Good Foundations: Towards a low carbon, high well-being built environment*, New Economics Foundation: London.

Aknin, L., Barrington-Leigh, C.P., Dunn, E., Helliwell, J., Biswas-Diener, R., Kemeza, I., et al. (2010). *Prosocial spending and well-being: Cross-cultural evidence for a psychological universal*, National Bureau of Economic Research: Cambridge, MA.

Amundadottir, M.L., Rockcastle, S., Khanie, M.S., Andersen, M. (2017). 'A human-centric approach to assess daylight in buildings for non-visual health potential, visual interest and gaze behavior', *Building and Environment*, 113, pp 5–21.

Anderson, J. (2014). 'Urban design and well-being', doctoral thesis, University of Cambridge.

Anderson, J. (2015). '"Living in a communal garden" associated with well-being while reducing urban sprawl by 40%: a mixed-methods cross-sectional study', *Frontiers in Public Health*, 3:173 (14 pages).

BSI (British Standards Institution) (2013). *Guidance on investigations for ground gas – Permanent gases and Volatile Organic Compounds (VOCs)*, BS 8576:2013, British Standards Institution: London.

Baker, N.V. (1987). 'Passive solar ventilation pre-heating', *Proceedings of 1987 European Conference on Architecture*, H.S. Stephens: Bedford.

Baker, N., Rassia, S. and Steemers, K. (2011). 'Designing for occupant movement in the workplace to improve health', *5th International Symposium on Sustainable Healthy Buildings*, Centre for Sustainable Healthy Buildings, Kyung Hee University: Seoul, pp 25–33.

Baker, N. and Standeven, M. (1996). 'Thermal comfort for free-running buildings', *Energy and Buildings*, pp 175–82.

Bauman, A. and Bull, F. (2007). *Environmental correlates of physical activity and walking in adults and children: A review of reviews*, National Centre for Physical Activity and Health, for the National Institute of Health and Clinical Excellence (NICE): Loughborough.

Bedrosian, T.A. and Nelson, R.J. (2013). 'Influence of the modern light environment on mood', *Molecular Psychiatry*, 18(7), pp 751–7.

Bluyssen, P. (2013). *The Healthy Indoor Environment*, Routledge: Abingdon.

Boubekri, M., Cheung, I.N., Reid, K.J., Wang, C.-H. and Zee, P.C. (2014). 'Impact of Windows and Daylight Exposure on Overall Health and Sleep Quality of Office Workers: A Case-Control Pilot Study', *Journal of Clinical Sleep Medicine*, 10(6), pp 603–11.

Brainard, G.C., Hanifin, J.R., Greeson, J.M., Byrne, B., Glickman, G., Gerner, E., Rollag, M.D. (2001). 'Action spectrum for melatonin regulation in humans: evidence for a novel circadian photoreceptor', *Journal of Neuroscience*, 21, pp 6405–12.

CABE (Commission for Architecture and the Built Environment) (2009). *Sustainable Places for Health and Well-being*. Commission for Architecture and the Built Environment: London.

Cattell, V., Dines, N., Gesler, W. and Curtis, S. (2008). 'Mingling, observing, and lingering: everyday public spaces and their implications for well-being and social relations', *Health Place*, pp 544–61.

Cawthorne, D. (1991). 'Buildings, lighting and the biological clock', *Building Technical File*, No. 32, pp 25–32.

Chalfont, G.E. (2008). *Design for Nature in Dementia Care*, Jessica Kingsley Publishers: London.

Chambers, R., Gullone, E. and Allen, N. (2009). Mindful emotion regulation: An integrative review, *Clinical Psychology Review*, pp 560–72.

Civil Aviation Authority (2016). *Aircraft Noise and Health Effects: Recent Findings*, CAP 1278, Civil Aviation Authority: West Sussex.

Crowther, D.R.G. and Baker, N.V. (1993). 'Breathing Walls: Facts and Fiction', *Proceedings of 'Solar Energy in Architecture and Planning'*, Florence, May 1993, H.S. Stephens and Associates: Bedford, pp 603–6.

Dalgard, O. and Tambs, K. (1997). 'Urban environment and mental health: A longitudinal study', *British Journal of Psychiatry*, pp 530–6.

de Dear, R. and Brager, G. (1998). 'Towards an adaptive model of thermal comfort and preference', *ASHREA Transactions*, pp 145–67.

D'Ippoliti, D. et al. (2010). 'The impact of heat waves on mortality in 9 European cities: results from the EuroHEAT project', *Environmental Health*, 9(37) (9 pages).

Dolan, P., Peasgood, T. and White, M. (2008). 'A review of the economic literature on the factors associated with subjective well-being', *Journal of Economic Psychology*, pp 94–122.

Dunn, E., Aknin, L. and Norton, M. (2008). 'Spending money on others promotes happiness', *Science*, pp 1687–8.

English Nature (2005). *Wildlife-friendly gardening - A general guide*, English Nature: Peterborough.

EPA (Environmental Protection Agency) (2009). *Residential Air Cleaners*, EPA 402-F-09-002, US Environmental Protection Agency: Washington, DC.

Evans, G. (2003). 'The Built Environment and Mental Health', *Journal of Urban Health*, pp 536–55.

Evans J.A. and Davidson, A.J. (2013). 'Health Consequences of Circadian Disruption in Humans and Animal Models', *Progress in Molecular Biology and Translational Science*, 119, pp 283–323.

Fanger, P. (1970). *Thermal comfort: Analysis and applications in environmental engineering*, Danish Technical Press: Copenhagen.

Fanger, P.O. (1982). *Thermal Comfort. Analysis and Applications in Environmental Engineering*, Krieger Publishing: Malabar, FL.

Foresight (2008). *Mental capital and well-being*, -The Government Office for Science: London.

Foster, R.G., Provencio, I., Hudson, D., Fiske, S., De Grip, W. and Menaker, M. (1991). 'Circadian photoreception in the retinally degenerate mouse (rd/rd)', *Journal of Comparative Physiology A*, 169(1), pp 39–50.

Greater London Authority (2008). *Improving Londoners' Access to Nature, London Plan*, GLA: London.

Guo, G. and Harris, K. (2000). 'The mechanisms mediating the effects of poverty on children's intellectual development', *Demography*, pp 431–47.

Gutman, L. and Akerman, R. (2008). *Determinants of aspiration*. Centre for Research on the Wider Benefits of Learning, Institute of Education: London.

Healy, J.D. (2003). 'Excess winter mortality in Europe: a cross country analysis identifying key risk factors', *Journal of Epidemiology and Community Health*, 57(10), pp 784–9.

Heathcote, E. (2006). *British Medical Journal*, 333, pp 1304–5.

Helliwell, J. and Putnam, R. (2004). 'The social context of well-being', *Philosophical Transactions of the Royal Society of London. Series B, Biological Sciences*, pp 1435–46.

Hofmann, S., Sawyer, A., Witt, A. and Oh, D. (2010). 'The effect of mindfulness-based therapy on anxiety and depression: A meta-analytic review', *Journal of Consulting and Clinical Psychology*, pp 169–83.

Honold, J., Wippert, P.-M. and van der Meer, E. (2014). 'Urban health resources: Physical and social constitutes of neighborhood social capital', *Procedia - Social and Behavioural Sciences*, pp 491–6.

Howard, E. (1989). *To-morrow: A Peaceful Path to Real Reform*, Swan Sonnenshein & Co.: London (subsequently published as *Garden Cities of To-morrow*, 1902).

Humphreys, M. (1978). 'Outdoor temperature and comfort indoors', *Building Research Practice*, 6(2), p 92.

Hunt, D.R.G. (1979). 'The Use of Artificial Lighting in Relation to Daylight Levels and Occupancy', *Building and Environment*, 14, pp 21–33.

Huppert, F. and So, T. (2013). 'Flourishing across Europe: Application of a new conceptual framework for defining well-being', *Social Indicators Research*, pp 837–61.

IPCC (Intergovernmental Panel on Climate Change) (1990). *Climate Change—The IPCC Impacts Assessment*, Australian Government Publishing Service: Canberra.

ISO (International Organization for Standardization) (2005). *7730:2005 - Ergonomics of the thermal environment*, International Organization for Standardization: Geneva.

Jenkins, A. (2011). 'Participation in learning and wellbeing among older adults', *International Journal of Lifelong Education*, pp 403–20.

Keller, R.C. (2015). *Fatal Isolation: The devastating Paris heat wave of 2003*, University of Chicago Press: Chicago.

Kent, T.S., McClure, L.A., Crosson, W.L., Arnett, D.K., Wadley, V.G. and Sathiakumar, N. (2009). 'Effect of sunlight on cognitive functioning of depressed and non-depressed participants', *Environmental Health*, 8(34) (14 pages).

Keon, W. J. and Pepin, L. (2009). *A healthy, productive Canada: A determinant of health approach*, Standing Senate Committee on Social Affairs, Science and Technology, the Subcommittee on Population and Health: Ottowa.

King, D., Thompson, P. and Darzi, A. (2014). 'Enhancing health and wellbeing though "behavioural design"', *Journal of the Royal Society of Medicine*, pp 336–7.

Kloog, I., Haim, A., Stevens, R.G. and Portnov, B.A. (2009). 'Global co-distribution of light at night (LAN) and cancers of prostate, colon, and lung in men', *Chronobiology International*, 26(1), pp 108–25.

Korte, C. and Kerr, N. (1974). 'Response to altruistic opportunities in urban and nonurban settings', *Social Psychology*, pp 183–4.

Krogh, J., Nordentoft, M., Sterne, J. and Lawlor, D. (2011). 'The effect of exercise in clinically depressed adults: systematic review and meta-analysis of randomized controlled trials', *Journal of Clinical Psychiatry*, pp 529–38.

Lafond, S., Charlesworth, A. and Roberts, A. (2016). *A perfect storm: an impossible climate for NHS providers' finances? An analysis of NHS finances and factors associated with financial performance*, Health Foundation: London.

Leaman, A. (1999). 'Window seat or aisle? Studies of buildings in use are showing the importance of personal control of one's environment, which includes windows', *Architects' Journal*, 4 March.

Leaman, A. and Bordass, B. (2007). 'Are users more tolerant of green buildings?' *Building Research and Information*, 35(6), pp 662–73.

Lee, I., Shiroma, E., Lobelo, F., Pushka, P., Blair, S. and Katzmarzyk, P. (2012). 'Impact of physical activity on the world's major non-communicable diseases', *Lancet*, pp 219–29.

Liedtke, C., Volker, S. and Knoop, M. (2013). 'The light direction and directional light – Towards a new quantification of an essential lighting quality criterion', *Proceedings of the CIE Centenary Conference, Paris, France*, CIE: Vienna, pp 542–51.

Lifetime Homes (2011). *Lifetime Homes Design Guide*, BRE Press: Watford.

Lund, H. (2002). 'Pedestrian environments and sense of community', *Journal of Planning Education and Research*, pp 301–12.

Marchant, J. (2016). *Cure: A journey into the science of mind over body*, Canongate: Edinburgh.

Marmot, M.G., Allen, J., Goldblatt, P., Boyce, T., McNeish, D., Grady, M. and Geddes, I. (2010). *Fair society, healthy lives: Strategic review of health inequalities in England post-2010*, The Marmot Review: London.

Marx, A., Fuhrer, U. and Hartig, H. (2000). 'Effects of classroom seating arrangements on children's question-asking', *Learning Environments Research*, pp 249–63.

Matias, L., Almeida, S., Pina Santos, C., Rebelo, M. and Correia Guedes M. (2009). 'Adaptive Thermal Comfort for Buildings in Portugal based on Occupants' Thermal Perception', *PLEA2009 - 26th Conference on Passive and Low Energy Architecture*, Quebec City.

Mehta, R. and Zhu, R. (2009). 'Blue or red? Exploring the effect of color on cognitive task performances', *Science*, pp 1226–9.

Meier, S. and Stutzer, A. (2008). 'Is volunteering rewarding in itself?' *Economica*, pp 39–59.

Meyers-Levy, J. and Zhu, R. (2007). 'The influence of ceiling height: The effect of priming on the type of processing that people use', *Journal of Consumer Research*, pp 174–86.

Moonen, P., Defraeye, T.W.J., Dorer, V., Blocken, B. and Carmeliet, J.E. (2012). 'Urban Physics: Effect of the micro-climate on comfort, health and energy demand', *Frontiers of Architectural Research*, 1(3), pp 197–228.

Murray, R., Jabbal, J., Thompson, J., Baird, B., Maguire, D. and Northern, E. (2016). *Quarterly monitoring report 21*, The King's Fund: London.

NICE (National Institute for Health and Clinical Excellence) (2008). *Promoting and creating built or natural environments that encourage and support physical activity*, National Institute for Health and Clinical Excellence: London.

Nicol, J. and Humphreys, M. (2002). 'Adaptive thermal comfort and sustainable thermal standards for buildings', *Energy and Buildings*, pp 563–72.

Nightingale, F. (1974) [First published 1859]. *Notes on Nursing: What it is and what it is not*, Blackie & Son Ltd: Glasgow and London.

Oliver, C.D., Nassar, N.T., Lippke, B.R. and McCarter, J.B. (2014). 'Carbon, Fossil Fuel, and Biodiversity Mitigation With Wood and Forests', *Journal of Sustainable Forestry*, 33:3, pp 248–75.

Oseland, N. (1995). 'Predicted and reported thermal sensations in climate chambers, offices and homes', *Energy in Buildings*, 23(2), pp 105–15.

Papamichael, K., Siminovitch, M., Veitch, J.A. and Whitehead, L. (2016). 'High Color Rendering Can Enable Better Vision without Requiring More Power', *LEUKOS*, 12(1–2), pp 27–38.

Parpairi, K. (1999). 'Daylighting in architecture – quality and user preference', PhD thesis, University of Cambridge.

Penckofer, S., Kouba, J., Byrn, M. and Estwing Ferrans, C. (2010). 'Vitamin D and depression: Where is the sunshine?' *Mental Health Nursing*, 31(6), pp 385–93.

Pettenkofer, M. (1858). *Über den Luftwechsel in Wohngebäuden* [Concerning Air Exchange in Residential Buildings], Der J.G. Cotta'schen Buchhandlung: Munich.

Plagnol, A. and Huppert, F. (2010). 'Happy to help? Exploring the factors associated with variations in rates of volunteering across Europe', *Social Indicators Research*, pp 157–76.

Provencio, I., Rodriguez, I.R., Jiang, G., Hayes W.P., Moreira, E.F. and Rollag, M.D. (2000). 'A Novel Human Opsin in the Inner Retina', *Journal of Neuroscience*, 20 (2), pp 600–5.

Roberts-Hughes, R. (2013). *City Health Check: How design can save lives and money*, RIBA Publishing: London.

Robertson, R., Wenzel, L., Thompson, J. and Charles, A. (2017). *Understanding NHS financial pressures*, The King's Fund: London.

Robson, E.R. (1874). *School Architecture*. John Murray: London.

Rybczynski, W. (1986). *Home – A short history of an idea*, Viking Penguin Books: Markham, Ontario.

Santamouris, M., Papanikolaou, N., Livada, I., Koronakis, I., Georgakis, C., Argiriou, A. and Assimakopoulos, D.N. (2001). 'On the impact of urban climate on the energy consumption of buildings', *Solar Energy*, 70(3), pp 201–16.

Schneider, M. (2002). *Do school facilities affect academic outcomes?* National Clearinghouse for Educational Facilities: Washington, DC.

Sempik, J., Rickhuss, C. and Beeston, A. (2014). 'The Effects of Social and Therapeutic Horticulture on Aspects of Social Behaviour', *British Journal of Occupational Therapy*, 77(6), pp 313–19.

Shahmohamadi, P., Che-Ani, A.I., Maulud, K.N.A., Tawil, N.M. and Abdullah, N.A.G. (2011). 'The Impact of Anthropogenic Heat on Formation of Urban Heat Island and Energy Consumption Balance', *Urban Studies Research*, Vol. 2011, Article ID 497524 (9 pages).

Shelter (2005). *Full house? How overcrowded housing affects families*, Shelter: London.

Sleep Council (2013). *The Great British Bedtime Report*, The Sleep Council: Skipton.

Sofi, F., Valecchi, D., Bacci, D., Abbate, R., Gensini, G., Casini, A., et al. (2011). 'Physical activity and risk of cognitive decline: a meta-analysis of prospective studies', *Journal of Internal Medicine*, pp 107–17.

Steemers, K. (2015). 'Architecture for well-being and health', *Daylight & Architecture*, Issue 23, pp 6–27.

Steemers, K., Baker, N., Crowther, D., Dubiel, J., Nikolopoulou, M. and Ratti, C. (1997). 'City Texture and Microclimate', *Urban Design Studies*, 3, pp 25–50.

Steinemann, A. (2009). 'Fragranced consumer products and undisclosed ingredients', *Environmental Impact Assessment Review*, 29(1), pp 32–8.

Steinemann, A. (2017). 'Ten questions concerning air fresheners and indoor built environments', *Building and Environment*, 111, pp 279–84.

Strife, S. and Downey, L. (2009). 'Childhood Development and Access to Nature', *Organization & Environment*, 22(1).

Tang, Y., Yang, L., Leve, L. and Harold, G.T. (2012). 'Improving executive function and its neurobiological mechanisms through a mindfulness-based intervention: Advances within the field of developmental neuroscience', *Child Development Perspectives*, pp 361–6.

Thaler, R. and Sunstein, C. (2008). *Nudge: Improving decisions about health, wealth and happiness*, Yale University Press: New Haven, CT.

Thompson Coon, J., Boddy, K., Stein, K., Whear, R., Barton, J. and Depledge, M. (2011). 'Does participating in physical activity in outdoor natural environments have a greater effect on physical and mental wellbeing than physical activity indoors? A systematic review', *Environmental Science & Technology*, pp 1761–72.

UKGBC (UK Green Building Council) (2016). *Health and Wellbeing in Homes*, UK Green Building Council: London.

Ulrich, R.S. (1984). 'View through a window may influence recovery from surgery', *Science*, 224(4647), pp 420–1.

Unwin, S. (2015). *Twenty-five buildings every architect should understand*, Routledge: Abingdon.

US DHHS (US Department of Health and Human Services) (2000). *Healthy people 2010: Understanding and improving health* (2nd edition), US Government Printing Office: Washington, DC.

Vartaniana, O., Navarrete, G., Chatterjee, A., Fich, L., Leder, H., Modrono, C., et al. (2013). 'Impact of contour on aesthetic judgments and approach-avoidance decisions in architecture', *PNAS (Proceedings of the National Academy of Sciences, USA)*, pp 10446–53.

Velux (2016). *Europeans on healthy living: the Healthy Homes Barometer 2016*, Velux Group: Hørsholm, Denmark.

Velux (2017). *Healthy Homes Barometer: Buildings and Their Impact on the Health of Europeans*, Velux Group: Hørsholm, Denmark.

Viitanen, H., Vinha, J., Salminen, K., Ojanen, T., Peuhkuri, R., Paajanen, L. and Lähdesmäki, K. (2010). 'Moisture and Bio-deterioration Risk of Building Materials and Structures', *Journal of Building Physics*, 33(3), pp 201–24.

Vitruvius, P. and Morgan, M. H. (1960). *Vitruvius: The ten books on architecture* (M. Morgan, trans.), Dover Publications: New York. [Original work published 1914.]

Ward Thompson, C. (2011). 'Linking landscape and health: The recurring theme', *Landscape and Urban Planning*, pp 187–95.

Washan, P., Stenning, J. and Goodman, M. (2014). *Building the Future: The economic and fiscal impacts of making homes energy efficient*, Verco/Cambridge Econometrics: Cambridge.

Weinhold, D. (2015). *Sick of noise: the health effects of loud neighbours and urban din*, Grantham Research Institute on Climate Change and the Environment, Working Paper No. 213, LSE: London.

Weinziri, J. (1914). 'The germicidal action of sunlight upon bacterial spores', *American Journal of Public Health*, 4(11), pp 969–74.

Wells, D. (2011). 'The value of pets for human health', *Psychologist*, 24, pp 172–6.

WHO (World Health Organization) (2009). *WHO guidelines for indoor air quality: dampness and mould*, World Health Organization Regional Office for Europe: Copenhagen.

WHO (2018). *WHO Housing and health guidelines*, World Health Organization: Geneva.

Wilson, S., Oliver, S., Mallet, H., Hutchings, H. and Card, G. (2007). *Assessing risks posed by hazardous ground gases to buildings*, CIRIA (Construction Industry Research and Information Association): London.

Xia, Q., Niu, J. and Liu, X. (2014). 'Dispersion of air pollutants around buildings: A review of past studies and their methodologies', *Indoor and Built Environment*, 23(2), pp 201–24.

Zhong, L., Goldberg, M.S., Gao, Y.T. and Jin, F. (1999). 'Lung cancer and indoor air pollution arising from Chinese-style cooking among nonsmoking women living in Shanghai, China', *Epidemiology*, 10(5), pp 488–94.

Endnotes

Unless otherwise stated, all URLs accessed April 2019

Prelims

1 Velux (November 2013). *Circadian House: Principles and Guidelines for Healthy*. Velux A/S: Hørsholm, Denmark. Available from http://thedaylightsite.com/wp-content/uploads/2017/10/Circadian-House-Principles-and-guidelines-for-healthy-homes.pdf

2 Chapter 2 is based in part on the following paper and re-appropriated here with the kind permission of the journal: Steemers, K. (2015). 'Architecture for well-being and health', Daylight & Architecture, Spring 2015, Issue 23, 6-27.

Chapter 1

1 A report of the Maison Air et Lumière can be found at www.velux.com/innovation/demo-buildings/maison-air-et-lumiere

2 The WHO definition of health is in its constitution, which was adopted when it was founded in 1948.

3 World Health Organization (WHO) (2018). *WHO Housing and health guidelines*. World Health Organization: Geneva.

4 Marchant, J. (2016). *Cure: A journey into the science of mind over body*. Canongate: Edinburgh.

5 Vitruvius, P., and Morgan, M.H. (1960). V*itruvius: The ten books on architecture.* (M. Morgan, trans.) Dover Publications: New York [Original work published 1914], p 10.

6 Marchant, *op. cit.*

7 *Ibid.*, p 294.

8 Heathcote, E. (2006). *British Medical Journal*, 333: 1304-1305.

9 Adams, A. (1996), *Architecture in the Family Way: Doctors, Houses, and Women, 1870-1900.* McGill-Queen's University Press: Montreal.

10 *Ibid.*, p 37.

11 *Ibid.*, p 39.

12 Campbell, M. (2005). 'What Tuberculosis did for Modernism: The Influence of a Curative Environment on Modernist Design and Architecture', *Medical History* 49(4), pp 463–88.

13 For further information on tuberculosis information from WHO (2018), see http://www.who.int/en/news-room/fact-sheets/detail/tuberculosis

14 Pettenkofer, M. (1858). *Über den Luftwechsel in Wohngebäuden*. Der J.G. Cotta'schen Buchhandlung: Munich.

15 Nightingale, Florence (1974) [First published 1859]. *Notes on Nursing: What it is and what it is not*. Blackie & Son Ltd: Glasgow and London.

16 Howard, E. (1989). *To-morrow: A Peaceful Path to Real Reform*. Swan Sonnenshein & Co.: London (subsequently published as *Garden Cities of To-morrow*, 1902).

17 Campbell, *op. cit.*, pp 474-475.

18 www.weissenhof2002.de/english/weissenhof.html

19 Campbell, *op. cit.*

20 Overdijk, M. (2015). *Richard Neutra's Therapeutic Architecture*, https://failedarchitecture.com/richard-neutras-therapeutic-architecture

21 Brainard, G.C., Hanifin, J.R., Greeson, J.M., Byrne, B., Glickman, G., Gerner, E. and Rollag, M.D. (2001). 'Action spectrum for melatonin regulation in humans: evidence for a novel circadian photoreceptor', *Journal of Neuroscience*, 21, pp 6405–12.

22 Foster, R.G., Provencio, I., Hudson, D., Fiske, S., De Grip, W. and Menaker, M. (1991). 'Circadian photoreception in the retinally degenerate mouse (rd/rd)', *Journal of Comparative Physiology A*, 169(1), pp 39–50.

23 Provencio, I., Rodriguez, I.R., Jiang, G., Hayes W.P., Moreira, E.F. and Rollag, M.D. (2000). 'A Novel Human Opsin in the Inner Retina', *Journal of Neuroscience*, 20 (2), pp 600–5.

24 Evans, J.A. and Davidson, A.J. (2013). 'Health Consequences of Circadian Disruption in Humans and Animal Models', *Progress in Molecular Biology and Translational Science*, 119, pp 283–323.

25 Amundadottir, M.L., Rockcastle, S., Khanie, M.S. and Andersen, M. (2017). 'A human-centric approach to assess daylight in buildings for non-visual health potential, visual interest and gaze behavior', *Building and Environment*, 113, pp 5–21.

26 Velux (2017). *Healthy Homes Barometer: Buildings and Their Impact on the Health of Europeans.* Velux Group: Horshølm, Denmark. http://velcdn.azureedge.net/~/media/com/health/healthy-home-barometer/507505-01%20barometer_2017.pdf

27 Velux (2016). *Europeans on healthy living: the Healthy*

Homes Barometer 2016. Velux Group: Hørsholm, Denmark, p 12. http://www.velux.com/article/2016/europeans-on-healthy-living-the-healthy-homes-barometer-2016
28 Velux (2017), *op. cit*.
29 UKGBC (July 2016). *Health and Wellbeing in Homes*. UK Green Building Council: London.
30 For more information about the BRE Home Quality Mark, see https://www.homequalitymark.com/what-is-the-hqm
31 Further details of the WELL Standard can be found at www.wellcertified.com/about-iwbi
32 See, for example, NHS, *Today's NHS – our current challenges*, My Health London, https://www.nuffieldtrust.org.uk/resource/nhs-hospitals-under-pressure-trends-in-acute-activity-up-to-2022
33 Lafond, S., Charlesworth, A. and Roberts, A. (2016). *A perfect storm: an impossible climate for NHS providers' finances? An analysis of NHS finances and factors associated with financial performance*. London: Health Foundation.
34 Murray, R., Jabbal, J., Thompson, J., Baird, B., Maguire, D. and Northern, E. (2016). *Quarterly monitoring report 21*. The King's Fund: London.
35 Marmot, M.G., Allen, J., Goldblatt, P., Boyce, T., McNeish, D., Grady, M. and Geddes, I. (2010). *Fair society, healthy lives: Strategic review of health inequalities in England post-2010*. The Marmot Review: London, p 18.
36 Friedman, D. (2010). 'Social impact of poor housing', Ecotec: London. http://southdevonrural.co.uk/userfiles/file/JC-JC13-Social-impact-of-poor-housing.pdf
37 Age UK (2012). 'The cost of cold: Why we need to protect the health of older people in winter'. Age UK: London. http://www.ageuk.org.uk/Documents/EN-GB/Campaigns/The_cost_of_cold_2012.pdf?dtrk=true
38 Washan, P., Stenning, J. and Goodman, M. (2014). 'Building the Future: The economic and fiscal impacts of making homes energy efficient', Verco/Cambridge Econometrics: Cambridge. http://www.energybillrevolution.org/wp-content/uploads/2014/10/Building-the-Future-The-Economic-and-Fiscal-impacts-of-making-homes-energy-efficient.pdf
39 Velux (2016), *op. cit.*, p 13
40 Robertson, R., Wenzel, L., Thompson, J. and Charles, A. (2017). *Understanding NHS financial pressures*. The King's Fund: London.
41 Marmot, et al., *op. cit*.
42 *Ibid.*, p 34.
43 *Ibid.*, pp 79–80.
44 UKGBC, *op. cit*.
45 Thaler, R. and Sunstein, C. (2008). *Nudge: Improving decisions about health, wealth and happiness*. Yale University Press: New Haven, CT.
46 King, D., Thompson, P. and Darzi, A. (2014). 'Enhancing health and wellbeing though "behavioural design"', *Journal of the Royal Society of Medicine*, pp 336–7.

Chapter 2

1 Vitruvius and Morgan, *op. cit*.
2 ISO (2005). *7730:2005 – Ergonomics of the thermal environment*. International Organization for Standardization: Geneva.
3 Bluyssen, P. (2013). *The Healthy Indoor Environment*. Routledge: Abingdon.
4 Fanger, P. (1970). *Thermal comfort: Analysis and applications in environmental engineering*. Danish Technical Press: Copenhagen.
5 Adaptive-comfort theory suggests that a human connection to the outdoors and control over the immediate environment allows building users to adapt to – and possibly even prefer – a wider range of thermal conditions than is generally considered comfortable.
- de Dear, R. and Brager, G. (1998). 'Towards an adaptive model of thermal comfort and preference', *ASHREA Transactions*, pp 145–67.
- Nicol, J. and Humphreys, M. (2002). 'Adaptive thermal comfort and sustainable thermal standards for buildings', *Energy and Buildings*, pp 563–72.
- Baker, N. and Standeven, M. (1996). 'Thermal comfort for free-running buildings', *Energy and Buildings*, pp 175–82.
6 Huppert, F. and So, T. (2013). 'Flourishing across Europe: Application of a new conceptual framework for defining well-being', *Social Indicators Research*, pp 837–61.
7 – Anderson, J. (2014). 'Urban design and well-being', doctoral thesis, University of Cambridge.

- Aked, J., Michaelson, J. and Steuer, N. (2010). *Good foundations: Towards a low carbon, high well-being built environment*. New Economics Foundation: London.
8 Roberts-Hughes, R. (2013). *City Health Check: How design can save lives and money*. RIBA Publishing: London.
9 CABE (2009). *Sustainable places for health and Well-being*. Commission for Architecture and the Built Environment: London.
10 NICE (2008). *Promoting and creating built or natural environements that encourage and support physical acivity*. National Institute for Health and Clinical Excellence.: London.
11 Dalgard, O. and Tambs, K. (1997). 'Urban environment and mental health: A longitudinal study', *British Journal of Psychiatry*, pp 530–6.
12 Evans, G. (2003). 'The Built Environment and Mental Health', *Journal of Urban Health*, pp 536–55.
13 Foresight (2008). *Mental capital and well-being*. The Government Office for Science: London.
14 Aked, J., Thompson, S., Marks, N. and Cordon, C. (2008). 'Five ways to well-being: the evidence', paper prepared for the 'Foresight' Project and available through www.foresight.gov.uk
15 'Large-scale' here denotes large sample sizes in number and/or over time; 'meta-studies' refers to 'studies of studies' – i.e. the compiling of multiple studies in order to look for consistent trends and findings.
16- Foresight, *op. cit.* - Dolan, P., Peasgood, T. and White, M. (2008). 'A review of the economic literature on the factors associated with subjective well-being', *Journal of Economic Psychology*, pp 94–122.
- Helliwell, J. and Putnam, R. (2004). 'The social context of well-being', *Philosophical Transactions of the Royal Society of London. Series B, Biological Sciences*, pp 1435–46.
17 'Global' here refers to a study or multiple studies across different regions; 'meta-studies' denotes the compilation of studies, potentially from different regions, looking for consistencies in findings.
18 - Krogh, J., Nordentoft, M., Sterne, J. and Lawlor, D. (2011). 'The effect of exercise in clinically depressed adults: systematic review and meta-analysis of randomized controlled trials', *Journal of Clinical Psychiatry*, pp 529–38.
- Lee, I., Shiroma, E., Lobelo, F., Pushka, P., Blair, S. and Katzmarzyk, P. (2012). 'Impact of physical activity on the world's major non-communicable diseases', *Lancet*, pp 219–29.
- Sofi, F., Valecchi, D., Bacci, D., Abbate, R., Gensini, G., Casini, A., et al. (2011). 'Physical activity and risk of cognitive decline: a meta-analysis of prospective studies', *Journal of Internal Medicine*, pp 107–17.
19 - Chambers, R., Gullone, E. and Allen, N. (2009). 'Mindful emotion regulation: An integrative review', *Clinical Psychology Review*, pp 560–72.
- Hofmann, S., Sawyer, A., Witt, A. and Oh, D. (2010). 'The effect of mindfulness-based therapy on anxiety and depression: A meta-analytic review', *Journal of Consulting and Clinical Psychology*, pp 169–83.
- Tang, Y., Yang, L., Leve, L. and Harold, G.T. (2012). 'Improving executive function and its neurobiological mechanisms through a mindfulness-based intervention: Advances within the field of developmental neuroscience', *Child Development Perspectives*, pp 361–6.
20 Gutman, L. and Akerman, R. (2008). *Determinants of aspiration*. Centre for Research on the Wider Benefits of Learning, Institute of Education: London.
21 Jenkins, A. (2011). 'Participation in learning and wellbeing among older adults', *International Journal of Lifelong Education*, pp 403–20.
22 - Aknin, L., Barrington-Leigh, C.P., Dunn, E., Helliwell, J., Biswas-Diener, R., Kemeza, I., et al. (2010). *Prosocial spending and well-being: Cross-cultural evidence for a psychological universal*. National Bureau of Economic Research: Cambridge, MA.
- Dunn, E., Aknin, L. and Norton, M. (2008). 'Spending money on others promotes happiness', *Science*, pp 1687–8.
23 - Plagnol, A. and Huppert, F. (2010). 'Happy to help? Exploring the factors associated with variations in rates of volunteering across Europe', *Social Indicators Research*, pp 157–76.
- Meier, S. and Stutzer, A. (2008). 'Is volunteering rewarding in itself?' *Economica*, pp 39–59.
24 Cattell, V., Dines, N., Gesler, W. and Curtis, S. (2008). 'Mingling, observing, and lingering: everyday public spaces and their implications for well-being and social

relations', *Health Place*, pp 544–61.
25 Lund, H. (2002). Pedestrian environments and sense of community. *Journal of Planning Education and Research*, 301-312.
26 Ward Thompson, C. (2011). 'Linking landscape and health: The recurring theme', *Landscape and Urban Planning*, pp 187–95.
27 Cattell, et al., *op. cit.*, p 552.
28 Bauman, A. and Bull, F. (2007). *Environmental correlates of physical activity and walking in adults and children: A review of reviews*. National Centre for Physical Activity and Health, for the National Institute of Health and Clinical Excellence (NICE): Loughborough.
29 Thompson Coon, J., Boddy, K., Stein, K., Whear, R., Barton, J. and Depledge, M. (2011). 'Does participating in physical activity in outdoor natural environments have a greater effect on physical and mental wellbeing than physical activity indoors? A systematic review', *Environmental Science & Technology*, pp 1761–72.
30 Anderson (2014), *op. cit.*
31 Guo, G. and Harris, K. (2000). 'The mechanisms mediating the effects of poverty on children's intellectual development', *Demography*, pp 431–47.
32 Marx, A., Fuhrer, U. and Hartig, H. (2000). 'Effects of classroom seating arrangements on children's question-asking', *Learning Environments Research*, pp 249–63.
33 Schneider, M. (2002). *Do school facilities affect academic outcomes?* National Clearinghouse for Educational Facilities: Washington, DC.
34 Honold, J., Wippert, P.-M. and van der Meer, E. (2014). 'Urban health resources: Physical and social constitutes of neighborhood social capital', *Procedia - Social and Behavioural Sciences*, pp 491–6.
35 Korte, C. and Kerr, N. (1974). 'Response to altruistic opportunities in urban and nonurban settings', *Social Psychology*, pp 183–4.
36 Anderson (2014), *op. cit.*
37 Exercise experts measure activity in metabolic equivalents: 1 MET is the energy it takes to sit quietly; vigorous activities burn more than 6 METs.
38 US DHHS. (2000). *Healthy people 2010: Understanding and improving health* (2nd edition.) US Government Printing Office: Washington, DC.
39 Baker, N., Rassia, S. and Steemers, K. (2011). 'Desiging for occupant movement in the workplace to improve health', *5th International Symposium on Sustainable Healthy Buildings*. Centre for Sustainable Healthy Buildings, Kyung Hee Universit: Seoul, pp 25–33.
40 UK Building Regulations (2010). *Approved Document M: Access to and use of buildings, Volume 1: Dwellings* (2015 edition incorporating 2016 amendments). NBS/RIBA Enterprises Ltd: Newcastle upon Tyne, p 8, para 1.17. This document defines principal storey as 'the floor level (of the dwelling) on which the main living space is located, where this is not the entrance storey' p. 52.
41 Lifetime Homes (2011). *Lifetime Homes Design Guide*. BRE Press: Watford.
42 These heights are specified in the UK Building Regulations, *Approved Document M*, p 49, para 3.44.
43 Mehta, R. and Zhu, R. (2009). 'Blue or red? Exploring the fffect of color on cognitive task performances', *Science*, pp 1226–9.
44 Meyers-Levy, J. and Zhu, R. (2007). 'The influence of ceiling height: The effect of priming on the type of processing that people use', *Journal of Consumer Research*, pp 174–186.
45 Vartaniana, O., Navarrete, G., Chatterjee, A., Fich, L., Leder, H., Modrono, C., et al. (2013). 'Impact of contour on aesthetic judgments and approach-avoidance decisions in architecture', *PNAS (Proceedings of the National Academy of Sciences, USA)*, pp 10446–53.
46 Keon, W.J. and Pepin, L. (2009). *A healthy, productive Canada: A determinant of health approach*. Standing Senate Committee on Social Affairs, Science and Technology, the Subcommittee on Population and Health: Ottowa.
47 Shelter (2005). *Full house? How overcrowded housing affects families*. Shelter: London.
48 *Ibid.*, p 19.
49 The 'Bedroom Standard' in Section 325 of the UK Housing Act 1985 requires a separate bedroom (of no less than 50 square feet) to be allocated to: (a) a person living together with another as husband and wife (whether that other person is of the same sex or the opposite sex); (b) a person aged 21 years or more; (c) two persons of the same sex aged 10 years to 20 years; (d) two persons (whether of the same sex or not) aged less than 10 years; (e) two persons of the same sex where

one person is aged between 10 years and 20 years and the other is aged less than 10 years; (f) any person aged under 21 years in any case where he or she cannot be paired with another occupier of the dwelling so as to fall within (c), (d) or (e) above.
50 Shelter, *op. cit*.
51 Unwin, S. (2015). *Twenty-five buildings every architect should understand*. Routledge: Abingdon, pp 255-264.

Chapter 3

1 Strife S and Downey L (March 2009). 'Childhood Development and Access to Nature', *Organization & Environment*, 22(1).
2 Chalfont, G.E. (2008). *Design for Nature in Dementia Care*. Jessica Kingsley Publishers: London.
3 Greater London Authority (2008). *Improving Londoners' Access to Nature, London Plan*. GLA: London.
4 English Nature (2005). *Wildlife-friendly gardening - A general guide*. English Nature: Peterborough.
5 Sempik, J., Rickhuss, C. and Beeston, A. (2014). 'The Effects of Social and Therapeutic Horticulture on Aspects of Social Behaviour', *British Journal of Occupational Therapy*, 77(6), pp 313–19.
6 Wells, D. (2011). 'The value of pets for human health', *Psychologist*, 24, pp 172–6.
7 Chalmers University of Technology (1 July 2014). 'Traffic noise is dangerous for your health: Solutions exist for dense cities', *ScienceDaily*. www.sciencedaily.com/releases/2014/07/140701085328.htm
8 Civil Aviation Authority (2016). *Aircraft Noise and Health Effects: Recent Findings*. CAP 1278. Civil Aviation Authority: West Sussex.
9 *Noise Action Plan Contours for Heathrow Airport 2014*. CAA Environmental Research and Consultancy Dept. Report No. 1504. Civil Aviation Authority: West Sussex, p 29.
10 Weinhold, D. (2015). *Sick of noise: the health effects of loud neighbours and urban din*. Grantham Research Institute on Climate Change and the Environment, Working Paper No. 213. LSE: London.
11 Moonen, P., Defraeye, T.W.J., Dorer, V., Blocken, B. and Carmeliet, J.E. (2012). 'Urban Physics: Effect of the micro-climate on comfort, health and energy demand', *Frontiers of Architectural Research*, 1(3), pp 197–228.
12 IPCC (1990). *Climate Change—The IPCC Impacts Assessment*. Australian Government Publishing Service: Canberra.
13 Santamouris, M., Papanikolaou, N., Livada, I., Koronakis, I., Georgakis, C., Argiriou, A. and Assimakopoulos, D.N. (2001). 'On the impact of urban climate on the energy consumption of buildings', *Solar Energy*, 70(3), pp 201–16.
14 Steemers, K., Baker, N., Crowther, D., Dubiel, J., Nikolopoulou, M. and Ratti C. (1997). 'City Texture and Microclimate', *Urban Design Studies*, 3, pp 25–50.
15 European Environment Agency (EEA). *Percentage of green urban areas in EU-27 core cities*, https://www.eea.europa.eu/data-and-maps/figures/percentage-of-green-urban-areas-1 2014
16 Shahmohamadi, P., Che-Ani, A.I., Maulud, K.N.A., Tawil, N.M. and Abdullah, N.A.G. (2011). 'The Impact of Anthropogenic Heat on Formation of Urban Heat Island and Energy Consumption Balance', *Urban Studies Research*, Vol. 2011, Article ID 497524, 9 pages.
17 The modelling of fluid-flow problems using Computational Fluid Dynamics.
18 Air-quality contour plots generated by Cambridge Environmental Research Consultants (CERC) using the ADMS-Urban model; plots included within the FAIRMODE (Forum for air-quality modelling in Europe) European Composite Map platform.

Chapter 4

1 Hammond, G. and Jones, C. (2011).*Inventory of Carbon & Energy (ICE) Version 2.0*, Sustainable Energy Research Team (SERT), Department of Mechanical Engineering, University of Bath.
2 Oliver, C.D., Nassar, N.T., Lippke, B.R. and McCarter, J.B. (2014). 'Carbon, Fossil Fuel, and Biodiversity Mitigation With Wood and Forests', *Journal of Sustainable Forestry*, 33:3, pp 248–75.
3 http://ecdb.wrap.org.uk
4 The International Commission on Illumination (Commission internationale de l'éclairage).
5 Hunt, D.R.G. (1979). 'The Use of Artificial Lighting in Relation to Daylight Levels and Occupancy', *Building and Environment*, 14, pp 21–33.
6 Air flow per hour measured in units of the whole building volume.

7 Leaman. A. (1999). Window seat or aisle? Studies of buildings in use are showing the importance of personal control of one's environment, which includes windows. *Architects' Journal*, 4 March.
8 www.passivhaustrust.org.uk/what_is_passivhaus.php
9 Viitanen, H., Vinha, J., Salminen, K., Ojanen, T., Peuhkuri, R., Paajanen, L. and Lähdesmäki, K. (2010). 'Moisture and Bio-deterioration Risk of Building Materials and Structures', *Journal of Building Physics*, 33(3), pp 201–24.
10 Crowther, D.R.G., Baker, N.V. (1993). 'Breathing Walls: Facts and Fiction', *Proceedings of international conference 'Solar Energy in Architecture and Planning', Florence, May 1993*. H.S. Stephens and Associates: Bedford, pp 603–6.
11 A building's 'free-running' temperature is that attained without the use of heating or cooling (see chapter 5).
12 SRI is defined as the difference in sound level (in dB) as sound travels through a partition. Comprehensive data for various constructions can be found in acoustic databases.
13 This simple detail can be applied to an existing floor in refurbishment situations. In new-build cases, more complex constructions can give higher performances. Refer to specialist manufacturer's details.

Chapter 5

1 Fanger, *op. cit.*
2 Humphreys, M. (1978). 'Outdoor temperature and comfort indoors', *Building Research Practice*, 6(2), p 92.
3 Matias, L., Almeida, S., Pina Santos, C., Rebelo, M. and Correia Guedes M. (2009). 'Adaptive Thermal Comfort for Buildings in Portugal based on Occupants' Thermal Perception', *PLEA2009* - 26th Conference on Passive and Low Energy Architecture, Quebec City.
4 An observational research method in which data is gathered for the same subjects repeatedly over a period of time.
5 Oseland, N. (1995). 'Predicted and reported thermal sensations in climate chambers, offices and homes', *Energy in Buildings*, 23(2), pp 105–15.
6 Parpairi, K. (1999). 'Daylighting in architecture – quality and user preference', PhD thesis, University of Cambridge.
7 Leaman, A. and Bordass, B. (2007). 'Are users more tolerant of green buildings?' *Building Research and Information*, 35(6), pp 662–73.
8 Defined as the ratio of the overall satisfaction score to the average score of individual environmental-performance indicators.
9 http://www.ons.gov.uk/ons/rel/hsq/health-statistics-quarterly/no--20--winter-2003/estimating-excess-winter-mortality-in-england-and-wales.pdf
10 Healy, J.D. (2003). 'Excess winter mortality in Europe: a cross country analysis identifying key risk factors', *Journal of Epidemiology and Community Health*, 57(10), pp 784–9.
11 D'Ippoliti, D., et al. (2010). 'The impact of heat waves on mortality in 9 European cities: results from the EuroHEAT project', *Environmental Health*, 9(37), 9 pages.
12 People found suffering from hypo- or hyperthermia often claim that they don't feel cold or hot, and are thus not prompted to take adaptive action.
13 Keller, R.C. (2015). *Fatal Isolation: The devastating Paris heat wave of 2003*. University of Chicago Press: Chicago.
14 This states that the intensity of radiation **I** is given by **I** = **W/r²** where **W** is the power of the source and **r** is the distance from the source.
15 Rybczynski W. (1986). *Home – A short history of an idea*. Viking Penguin Books: Markham, Ontario.
16 Baker, N.V. (1987). 'Passive solar ventilation pre-heating', *Proceedings of 1987 European Conference on Architecture*. H.S. Stephens: Bedford.
17 This can be a very satisfactory solution in a well-controlled predominantly naturally ventilated building where the mechanical backup is controlled by detecting CO_2 levels, a good indicator of IAQ. The mechanical equipment can be located in areas that are part of the passive system, such as stacks or roof ventilators.
18 See LEED material-ingredient reporting: https://www.usgbc.org/credits/new-construction-core-and-shell-schools-new-construction-retail-new-construction-healthca-35?view=language
19 For information on toxic materials in the WELL Standard, see: https://standard.wellcertified.com/air/toxic-material-reduction
20 A list of the worst materials prevalent in the building industry, and to be avoided, is available here: https://living-future.org/declare/declare-about/red-list/
21 Zhong, L., Goldberg, M.S., Gao, Y.T. and Jin, F. (1999).

'Lung cancer and indoor air pollution arising from Chinese-style cooking among nonsmoking women living in Shanghai, China', *Epidemiology*, 10(5), pp 488–94.
22 Steinemann, A. (2009). 'Fragranced consumer products and undisclosed ingredients', *Environmental Impact Assessment Review*, 29(1), pp 32–8.
23 Steinemann, A. (2017). 'Ten questions concerning air fresheners and indoor built environments', *Building and Environment*, 111, pp 279–84.
24 Tiny particles or droplets in the air that are 2.5 microns or less in width.
25 Xia, Q., Niu, J. and Liu, X. (2014). 'Dispersion of air pollutants around buildings: A review of past studies and their methodologies', *Indoor and Built Environment*, 23(2), pp 201–24.
26 Environmental Protection Agency (EPA) (2009). *Residential Air Cleaners*, EPA 402-F-09-002. US Environmental Protection Agency: Washington, DC.
27 For example, an interactive map showing radon-affected areas in the UK is available from Public Health England's website: www.ukradon.org/information/ukmaps
28 British Standards Institution (BSI) (2013). *Guidance on investigations for ground gas – Permanent gases and Volatile Organic Compounds (VOCs)*, BS 8576:2013. British Standard Institution: London.
29 http://oer2go.org/mods/en-boundless/www.boundless.com/chemistry/textbooks/boundless-chemistry-textbook/chemistry-and-the-real-world-25/pollution-and-environmental-issues-175/indoor-pollution-radon-666-10529/index.html
30 Wilson, S., Oliver, S., Mallet, H., Hutchings, H. and Card G. (2007). *Assessing risks posed by hazardous ground gases to buildings*. CIRIA (Construction Industry Research and Information Association): London.
31 WHO (2009). *WHO guidelines for indoor air quality: dampness and mould*. World Health Organization Regional Office for Europe: Copenhagen.
32 Individual wall-mounted mechanical ventilation units with heat recovery are also available.
33 www.passivhaus.org.uk/standard.jsp?id=122
34 UK Building Regulations (2010). *Approved Document J:Combustion appliances and fuel storage systems* (incorporating 2010 and 2013 amendments). NBS/RIBA Enterprises Ltd: Newcastle upon Tyne, p 41, para 2.34.
35 In this situation, it refers to a fan located in a parallel duct that can be bypassed when the system is in natural-ventilation mode to avoid the resistance to airflow of the stationary fan.
36 Robson E.R. (1874). *School Architecture*. John Murray: London, p 167.
37 www.who.int/uv/intersunprogramme/en
38 Boubekri, M., Cheung, I.N., Reid, K.J., Wang, C.-H. and Zee, P.C. (2014). 'Impact of Windows and Daylight Exposure on Overall Health and Sleep Quality of Office Workers: A Case-Control Pilot Study', *Journal of Clinical Sleep Medicine*, 10(6), pp 603–11.
39 CEN (2018). *Daylight of buildings*, Draft European Standard, FprEN 17037:2018 E. Committee for Standardization (CEN): Brussels.
40 In typical indoor clothing, direct sun through a window will create an apparent increase of temperature of between 3K and 6K.
41 Cawthorne, D. (1991). 'Buildings, lighting and the biological clock', *Building Technical File*, No. 32, pp 25–32.
42 Report describing initial Literature Review on Circadian Lighting, CIBSE/BRE Client Report, B137170-01 (Dated January 2017).
43 Bedrosian, T.A. and Nelson, R.J. (2013). 'Influence of the modern light environment on mood', *Molecular Psychiatry*, 18(7), pp 751–7.
44 Sleep Council (2013). *The Great British Bedtime Report*. The Sleep Council: Skipton.
45 Kloog, I., Haim, A., Stevens, R.G. and Portnov, B.A. (2009). 'Global co-distribution of light at night (LAN) and cancers of prostate, colon, and lung in men', *Chronobiology International*, 26(1), pp 108–25.
46 Ulrich, R.S. (1984). 'View through a window may influence recovery from surgery', *Science*, 224(4647), pp 420–1.
47 Nair, R. and Maseeh, A. (2012). 'Vitamin D: The "sunshine" vitamin', *Journal of Pharmacology & Pharmacotherapeutics*, 3(2), pp 118–26. DOI: 10.4103/0976-500X.95506; PMCID: PMC3356951.
48 www.cancerresearchuk.org/about-cancer/causes-of-cancer/sun-uv-and-cancer/how-does-the-sun-and-uv-cause-cancer
49 Weinziri, J. (1914). 'The germicidal action of sunlight

upon bacterial spores, *American Journal of Public Health*, 4(11), pp 969–74.
50 Kloog, et al., *op. cit.*
51 Bedrosian and Nelson, *op. cit.*
52 Sleep Council, *op. cit.*
53 Ulrich, *op. cit.*
54 Kent, T.S., McClure, L.A., Crosson, W.L., Arnett, D.K., Wadley, V.G. and Sathiakumar, N. (2009). 'Effect of sunlight on cognitive functioning of depressed and non-depressed participants', *Environmental Health*, 8(34), 14 pages.
55 Penckofer, S., Kouba, J., Byrn, M. and Estwing Ferrans, C. (2010). 'Vitamin D and depression: Where is the sunshine?' *Mental Health Nursing*, 31(6), pp 385–93.
56 Cawthorne, D. (1991). 'Buildings, lighting and the biological clock', *Building Technical File*, No. 32, pp 25–32.
57 Most objects are coloured because they absorb certain wavelengths due to specific pigments in their surface.
58 'nm' in the figure refers to the nanometre: a unit of length equal to one billionth of a metre (10^{-9}m).
59 Papamichael, K., Siminovitch, M., Veitch, J.A. and Whitehead, L. (2016). 'High Color Rendering Can Enable Better Vision without Requiring More Power', *LEUKOS*, 12(1–2), pp 27–38.
60 Liedtke, C., Volker, S. and Knoop, M. (2013). 'The light direction and directional light – Towards a new quantification of an essential lighting quality criterion', *Proceedings of the CIE Centenary Conference, Paris, France*. CIE: Vienna, pp 542–51.
61 sigmaluminous.com/why-dynamic-led-lighting-systems-for-schools-are-the-best

Chapter 6

1 Anderson, J. (2015). '"Living in a communal garden" associated with well-being while reducing urban sprawl by 40%: a mixed-methods cross-sectional study', *Frontiers in Public Health*, 3:173, 14 pages.
2 An interactive map showing radon-affected areas for the whole of the UK is available from Public Health England's website: www.ukradon.org/information/ukmaps
3 Phase Change Materials (PCM) can be used for thermal-energy storage, aiming to enhance building energy efficiency. Gypsum plasterboard incorporates PCM capsules that absorb heat energy during the day and releases them at night, reducing temperature fluctuations. The effect is similar to thermal mass but with a reduced physical weight.
4 Higher Education Statistics Agency (HESA) figures show that the number of students who disclosed a mental-health condition almost doubled between 2012 and 2015 to nearly 45,000, rising to 53,000 in 2016–17: https://www.hesa.ac.uk/data-and-analysis/students/table-15
5 https://www.theguardian.com/artanddesign/2013/aug/29/carbuncle-cup-student-housing-ucl
6 https://www.dezeen.com/2013/08/29/carbuncle-cup-2013-winner-announced/

Chapter 7

1 Nightingale, *op. cit.*

Index

Page numbers in *italic* indicate figures and in **bold** indicate tables.

Image Credits

Architectural Press Archive / RIBA Collections Figure 1.5

Ben Channon Figures 6.1, 6.7–8, 6.13–14, 6.25

CERC Figure 3.14

Civil Aviation Authority Figure 3.10

Douglas Cawthorne Figure 5.25

Feilden Clegg Bradley Studios Figure 6.2

IESVE Figure 5.23

Koen Steemers Figures 2.1, 6.10

Koen Steemers / Ben Channon Figures 6.23–24

Lasdun Archive / RIBA Collections Figure 6.22

Mole Architects / David Butler Figure 6.9

Nicholas Breach / RIBA Collections Figure 1.4

Nick Baker Figures 3.13, 4.2, 4.6, 4.22, 5.7–8, 5.10–13, 5.20, 5.24, 5.28–29, 6.5–6, 6.11–12, 6.15–17

Nick Baker / Ben Channon Figures 3.2–3, 3.8–9, 3.11–12, 4.4–5, 4.7–8, 4.11–14, 4.16–21, 4.23–4, 5.2–6, 5.9, 5.14–15, 5.18, 5.21–22, 5.26, 6.4, 6.18–21

Qubed Architecture Figure 5.19

RIBA Collections Figures 1.6–8

Traffic Noise, Greater London Council Figure 3.7

Velux A/S Figures 1.2–3, 1.9–10, 2.2–13, 3.1, 3.4–6, 4.1, 4.3, 4.9–10, 4.15, 5.1, 5.17, 5.27, 6.3, 7.1

Velux A/S / Adam Mørk Figure 1.1